A MATTER OF DEGREES

Gino Segrè

A MATTER OF DEGREES

What Temperature Reveals About the Past and Future of Our Species, Planet, and Universe

VIKING

VIKING
Published by the Penguin Group
Penguin Putnam Inc., 375 Hudson Street, New York, New York 10014, U.S.A.
Penguin Books Ltd, 80 Strand, London WC2R 0RL, England
Penguin Books Australia Ltd, 250 Camberwell Road, Camberwell, Victoria 3124, Australia
Penguin Books Canada Ltd, 10 Alcorn Avenue, Toronto, Ontario, Canada M4V 3B2
Penguin Books India (P) Ltd, 11 Community Centre, Panchsheel Park,
New Delhi – 110 017, India
Penguin Books (N.Z.) Ltd, Cnr Rosedale and Airborne Roads, Albany,
Auckland, New Zealand
Penguin Books (South Africa) (Pty) Ltd, 24 Sturdee Avenue,
Rosebank, Johannesburg 2196, South Africa

Penguin Books Ltd, Registered Offices: Harmondsworth, Middlesex, England

First published in 2002 by Viking Penguin, a member of Penguin Putnam Inc.

10 9 8 7 6 5 4 3 2 1

Grateful acknowledgment is made for permission to reprint the following copyrighted works:
"Fire and Ice" from *The Poetry of Robert Frost,* edited by Edward Connery Lathem. © 1969 by Henry Holt and Company. Reprinted by permission of Henry Holt & Co., LLC.
Excerpt from "The Quaker Graveyard in Nantucket" from *Lord Weary's Castle* by Robert Lowell. Copyright 1946 and renewed 1974 by Robert Lowell. Reprinted by permission of Harcourt, Inc.
"Cosmic Gall" from *Collected Poems 1953–1993* by John Updike. Copyright © 1993 by John Updike. Used by permission of Alfred A. Knopf, a division of Random House, Inc.

ILLUSTRATION CREDITS: Page 24: Richard Ellis; 55: Sezione di Zoologia "La Specola" del Museo di Storia Naturale dell'Università degli Studi di Firenze; 56: Istituto e Museo di Storia della Scienza; 69, 74, 89, 106, 114, 123, 134, 137, 167, 181: Felice Macera; 95: A. Geike, *The Life of Sir Roderick Murchison* (John Murray, London, 1875); 141: National Oceanic and Atmospheric Administration, Photo Library, National Undersea Research Program; 144: from V. Robigou, J. R. Delaney, and D. S. Stakes, Large massive sulfide deposits in a newly discovered active hydrothermal system, the high rise field, endeavor segment, Juan de Fuca Ridge, in Geophysics Research letters 20 (1993), courtesy Veronique Robigou, University of Washington School of Oceanography; 184: AIP (American Institute of Physics) Emilio Segrè Visual Archives; 195: Courtesy Sudbury Neutrino Observatory; 198: Photograph by C. E. Wynn-Williams, courtesy AIP Emilio Segrè Visual Archives; 219: Lucent Technologies' Bell Laboratories, courtesy AIP Emilio Segrè Visual Archives, Physics Today Collection; 228: AIP Emilio Segrè Archives, Chandrasekhar Collection; 243: Rijksmuseum Voor de Geschiedenis der Natuurwetenschappen te Leiden, courtesy AIP Emilio Segrè Visual Archives; 265: United States Patent and Trademark Office, courtesy of Gene Dannen; 270: Photo by Paul Ehrenfest, courtesy AIP Emilio Segrè Visual Archives

LIBRARY OF CONGRESS CATALOGING IN PUBLICATION DATA
Segrè, Gino.
A matter of degrees : what temperature reveals about the past and future
of our species, planet, and universe / Gino Segrè.
p. cm.
Includes bibliographical references and index.
ISBN 0-670-03101-1
1. Temperature measurements—Popular works. I. Title.
QC271 .S44 2001
536'.5'0287—dc21 2001046942

This book is printed on acid-free paper. ♾

Printed in the United States of America Set in Sabon Designed by Francesca Belanger

For Bettina

Contents

Introduction: The Ruler, the Clock, and the Thermometer

MOST OF US are likely to start our day with a series of questions: Where do I have to go? What time is it? How cold is it? In going to sleep, we anticipate tomorrow's answers to those same questions. The measurements of length, time, and temperature, implicit or explicit, set our life's rhythms. I'm particularly fascinated by temperature, the subtlest of the three. While new ideas expand our horizons, the everyday understanding of length and time has not changed appreciably in millennia. We've had rulers and clocks for a long time. This is not the case with temperature. Even though we can agree that a baby immediately knows hot from cold, our ability to measure temperature is only a few hundred years old. Our scientific understanding of even a gas's temperature—the average kinetic energy of molecules in thermal equilibrium—is much more recent.

Traditionally, science books intended for the general public describe a specific discipline or a particular problem. Books on cosmology or genetics or neuroscience are useful and often wonderful. I'm taking a different path, using the measurement of temperature as a guide in exploring many aspects of science. Such a wide sweep inevitably entails selection; the ensuing choices reflect my own background and taste, as well as my ignorance and knowledge. As a caveat, I should first tell you who I am and where this story is heading.

I'm a physicist. When people ask what I do for a living, I tell them I'm in the family business. My brother is a physicist, my nephew is one, lots of cousins are, my uncle received the Nobel Prize in physics, my wife's father was a well-known German physicist, and her sister is married to an even more famous Viennese physicist. Physics, my professional life, also has a familial side for me.

Two generations ago, the family business was paper. My Jewish grandfather Giuseppe moved, as a young man, from Mantova in northern Italy to Tivoli, a city about fifteen miles west of Rome. He built there a small paper mill that grew with the increased demands of the prospering capital of a newly unified Italy. The new and yet old country, once barely tolerant of Jewish enterprise, now encouraged it. Giuseppe was rewarded for his efforts when the new Italy awarded him the honorific title of *commendatore.*

Tivoli had thrived in Roman times. Known then as Tibur, it is nestled in the foothills of the Apennines, surrounded by poplar forests and cooled by the Aniene River with its numerous waterfalls. Tibur was an agreeable place to spend the hot summer months. Villas and temples sprung up as the wealth of the Roman Empire grew. In the second century A.D., Hadrian built his magnificent residence at the point where the Tibur hills met the Roman countryside. As described in Marguerite Yourcenar's *Memoirs of Hadrian,* it was more than just a villa. With its theater, reflecting pool, and other outlying buildings, it was probably the single largest residence of the classical world and yet it exuded quiet and tranquility. Yourcenar imagines the emperor thinking:

> Once more I have gone back to the Villa, to its garden pavilions built for privacy and for repose, to the vestiges of a luxury free of pomp, and as little imperial as possible,

> conceived of rather for the wealthy connoisseur who tries to combine the pleasures of art with the charms of rural life.

This jewel had been long abandoned, but excavations began in 1870 at Villa Adriana, as it was now called, just as Rome became the capital of a reborn Italy.

In the Renaissance, Tivoli, the new name for Tibur, was, as before, a fabled retreat from the heat of the nearby metropolis. In 1550 Cardinal Ippolito d'Este began the transformation of an old monastery into Villa d'Este, a sumptuous residence with Italy's finest display of Renaissance fountains. He built it on the hillside to heighten the effect of the cascading waters and also to allow the cardinals and the nobles, promenading along its cool paths, to glimpse the dome of Saint Peter's Basilica in the distance. Tivoli became a byword for elegance and charm. Such far-flung establishments as the amusement park in Copenhagen still carry its name.

The nineteenth century saw the arrival of industry. Paper mills require trees for wood pulp, plentiful water, power, and, hopefully, a nearby market. Tivoli had all of that, and so my grandfather built his mill just below Villa d'Este, on the site of the old Roman Temple of Hercules. The ruins of the temple provided the backbone of the factory, in what to present sensibility is an unthinkable sacrilege. But in those days, the growing demands of the new Rome made the rocks of old Rome seem an appropriate foundation. Late in his own life my father joked that the only trace of the Segrè family in Tivoli would be a plaque saying "Here is the Temple of Hercules, desecrated by the Segrè family, but restored to its former grandeur in———"

My grandparents had three children, all boys. They grew up in a world dramatically poised between the old and the new. The oldest, Angelo, my father, wandered as a child through the ruins of Villa Adriana, collected Roman coins, and studied the past. He eventually became a professor of ancient history, but he wanted to do more than just chronicle the past. He wanted to understand how people paid their bills, what they traded, the way their economy worked, how the currencies in the Mediterranean were valued, what Romans did in times of fiscal crisis. His most significant work was a two-volume treatise, *Metrology and Monetary Circulation in the Ancient World*, metrology being the study of measurement. I still remember him describing to me his excitement when, told of the discovery of an ancient storeroom full of broken pots, he realized he could predict the size each pot would have when reassembled. He knew what had been in the room, what was stored in the pots, who sold them, who bought them, and for how much. He knew all the measurements.

A charming, lovable, idiosyncratic man, impractical but immensely learned, my father increasingly came to regard the study of the ancient world as a kind of luxury. Fascinated by the emergence of quantum mechanics, relativity, genetics, and the idea of an expanding universe, he urged his children to study science, possibly regretting his own early decision not to do so. Perhaps another way of viewing my father's feelings is that historical awareness was so ingrained in him that he urged others to explore what to him was foreign.

The middle son, Marco, took the conventional route, staying in the old family business and running the paper mill. The measurements he studied were the prosaic but certainly important ones of balance sheets, cash flows, and growth curves.

In the mid-1920s, my grandfather's third son, Emilio, still an undergraduate at the University of Rome, began doing research with Enrico Fermi. Fermi, newly arrived in Rome, was only four years older than Emilio, but he was already a professor and beginning his rise to fame. In collaboration with Fermi and others, Emilio went on to a very successful career in physics, both in Europe and later in the United States.

Emilio is best known for his work with Fermi on neutrons and for the discovery of the antiproton, which led to his receiving the 1959 Nobel Prize in physics with Owen Chamberlain. I like to remember him for a less well known discovery: that of the element technetium and, in particular, the measurement of its half-life. The story goes as follows: Emilio became acquainted during a visit to Berkeley in 1937 with the great American experimentalist Ernest Lawrence, the builder of the first cyclotron. They corresponded regularly after that, since they had similar interests. At one point, Lawrence sent to Emilio, still in Italy, a molybdenum foil that had been placed in Lawrence's California cyclotron. Emilio suspected the cyclotron bombardment might have in fact led to the production of the forty-third element of the periodic table, an element that had never before been detected. After a careful set of chemical separations done with the help of his colleague Carlo Perrier, Emilio found he was right; they named the new element technetium. One of the reasons it had been missed in earlier chemical analyses was that technetium has several chemically identical forms, none of them stable.

I know the discovery had a special meaning for my uncle because, when World War II was over and Emilio could finally visit his father's grave, he brought some technetium with him. As he says,

> I scattered a small sample of technetium on my father's tomb at the Verano Cemetery in Rome, my tribute of love and respect as a son and as a physicist. The radioactivity

was minuscule, but its half-life of hundreds of thousands of years will last longer than any monument I could offer.

Emilio turned to history as he grew old. His first nonscientific venture was a biography of his mentor, Enrico Fermi. Later, trying to frame what he had seen in his life, he wrote a history of twentieth-century physics and finally a book on the history of classical or prequantum physics. These books are an attempt by Emilio to uncover, as he admitted, the meaning of Dante's phrase "Chi furono li maggiori tuoi?" (Literally, this means, "Who were your greaters?" but in an intellectual sense, "Who were your ancestors?")

My uncle's involvement and my father's guidance surely pushed me to participate in physics, the new family business. My father went a step further—he announced I should become a theoretical physicist. When I pressed him on how he reached this decision, he replied that theoretical physics seemed to be a profession with two cardinal virtues: you can tell right from wrong and you don't have to talk to anyone you don't want to talk to. Although both assertions are arguable, I became one anyway and certainly proved I was an obedient son. Over the past thirty years I have mainly worked on problems in elementary particles, occasionally branching out to condensed matter physics and astrophysics.

As I look back at my own pursuits and at the lives of my father and his brothers, I see us all drawn to the three types of measurement that rule our lives: length, time, and temperature. The volume of an amphora, the half-life of technetium, and the temperature in a neutron star are measured with sophisticated instruments. Simpler appraisals are made with the ruler, the clock, and the thermometer.

I knew at the outset that I wanted to incorporate in this book a discussion of some of the big questions science has addressed in the past century, many of which remain unanswered. In endeavoring to do this, I was pleased to discover that temperature was necessarily part of the narrative, not a peripheral marker. Consider three examples.

Our Earth was formed about 4.5 billion years ago from a protoplanetary disc, but when did life first appear? Although it was certainly present 3.7 billion years ago, was the intervening period, 800 million years, long enough for primordial organic molecules to assemble into genetic material? Was the necessary aquatic environment present? The answers depend on early Earth's temperature—how long a favorable climate existed and how resistant life was to thermal jumps. If conditions were such that life could not have formed that quickly on Earth, we must search for its origins elsewhere in the solar system. If life came from elsewhere, where did favorable conditions exist four billion years ago and how did life make the journey to Earth?

Next, consider the universe's birth in the cosmic explosion known as the Big Bang. Unimaginably hot in the beginning, the universe cooled over the course of 300,000 years to 5500 degrees Fahrenheit (3000 degrees Kelvin is the way this is usually presented). Experimental evidence says the temperature in that 5500-degree universe was almost completely uniform, the same at one point as at another. Yet it cannot have been completely uniform, or galaxies, stars, and planets would not have evolved. The signals from temperature fluctuations of less than a degree, present at that early time, are now studied with the tools of modern astronomy.

As a third example, consider the rather strange concept of a lowest possible temperature, an absolute zero. The notion of approaching that limit, first glimpsed less than 200 years ago, has turned into the exploration of a new world in which rules of quantum mechanics dictate behavior, wires have no electrical resistance, and flowing fluids experience no friction. This world, so remote from our own experience, has its counterpart in stellar interiors. Beyond that, it may yield important new technologies that can serve our everyday lives.

Some of temperature's most interesting puzzles, perhaps not as sweeping as the three just mentioned, are no less important. There is no simple answer to why our bodies maintain a constant temperature whether we live in the Arctic or the Sahara, why that temperature is 98.6 degrees, or why most mammals and birds have approximately the same temperature. The demand for unvarying brain readiness and response is clearly an important factor. But more is involved, as we see from the myriad adaptive mechanisms our animal brethren have adopted. Nor is there a complete answer to what advantages are offered us by the evolution of fever as a response to infection.

This book raises many puzzles. Some of the contents may seem paradoxical: for instance, it's surprising that we know the temperature at the center of the Sun with greater precision than that at the center of the Earth. However, many of the problems addressed have explanations that seem almost obvious upon reflection. While I don't claim to offer an overarching view of science, I stress the connections of the approaches as well as of the solutions. Temperature is the thread.

A MATTER OF DEGREES

Chapter 1

98.6

NINETY-EIGHT POINT SIX. It is extraordinary how alike we are; a thermometer under the tongue of an Inuit on an Arctic ice floe, a pygmy in the Ituri forest, or a stockbroker on the floor of the New York Stock Exchange gives the same reading. Yellow, black, brown, or white, tall or short, fat or thin, old or young, male or female, it's still 98.6. For a one-month-old baby, a twenty-year-old athlete, and a centenarian, body temperature is still the same. Muscles bulge or atrophy, teeth erupt or fall out, vision is acute or clouded by cataracts, heart rates double under stress, breathing fluctuates wildly, you shiver uncontrollably or sweat buckets, but temperature stays the same. And you feel sick if it varies by the merest 2 percent. If it rises or falls by much more than 5 percent, you should consider heading for the emergency room. The resemblance of one human to another in this respect is truly remarkable. Respiration, perspiration, excretion, and other bodily functions have huge swings, keyed to maintaining a constant body temperature.

Strictly speaking, 98.6 is only a convenient shorthand because measurements vary, albeit in a predictable way, over the range of our bodies. Our skin temperature is usually some 6 degrees lower than our internal temperature, as you can easily verify by placing a thermometer between your fingers rather than under your tongue. Oral and rectal readings differ as well, the latter commonly being about one degree higher than the former. Internal temperature varies between organs depending on metabolism and blood flow. Even before temperature was measured, our ancestors thought the hottest part of the body was the heart and in particular the heart of "hot-blooded types." Now, more prosaically, we have discovered that the distinctly less passionate liver, usually hovering at close to 105 degrees, has the honor of being the hottest.

The idea that all humans have the same body temperature would have seemed very curious before the seventeenth century. Only rough thermometers existed and none were used to make careful readings, comparing one human to another. The assumption was that body temperature, measured only externally and roughly at that, reflected the local climate and therefore was higher in the tropics than in temperate regions. The first problem posed in Johannes Hasler's influential *De logistica medica,* published in 1578, is "to find the natural degree of temperature of each man, as determined by his age, the time of the year, the elevation of the pole (i.e., latitude) and other influences." Hasler's elaborate tables indicated to doctors how they should mix their medicines according to degrees of hot and cold, the needs of the patient, and what his or her surroundings were like. Of course, the notion that fever was associated with sickness was well known to him. That's why he urged practitioners to be alert to temperature shifts.

We now know our body temperature doesn't vary with location. It does change slightly according to the time of day, gradually rising to a midafternoon peak typically 1.5 degrees higher than the nighttime minimum; 98.6 is simply a daily average. Yet even that statement needs to be qualified, as *Harrison's Principles of Internal Medicine* tells us:

> Whereas the "normal" temperature in humans has been said to be 98.6°F. on the basis of Wunderlich's original observations more than 120 years ago, the overall mean oral temperature for healthy individuals aged 18 to 40 years is actually 98.2°F.

A temperature reading above normal is called pyrexia, or, more commonly, fever, and one below normal is known as hypothermia. The regulatory mechanisms that keep us for the most part in the normal range are directed by a master control system embedded deep in the brain. The hypothalamus, the tiny organ that sets temperature, also controls the secretions that dictate many of the key metabolic functions, adjusts water, sugar, and fat levels in our body, and guides the release of hormones that both inhibit and enhance our activities. Harvey Cushing, the great early-twentieth-century American physician who studied the behavior of the hypothalamus and the pituitary glands, described our hypothalamus as follows:

> Here in this well concealed spot, almost to be covered with a thumbnail, lies the very mainspring of primitive existence—vegetative, emotional, reproductive—on which with more or less success, man has come to superimpose a cortex of inhibitions.

Our bodies generate heat by a variety of metabolic mechanisms fueled by food and drink. The body dissipates approximately 85 percent of the heat through the skin, the rest exiting by respiration and excretion. Since skin is the principal point of entry and exit for heat, that's where we should look for a connection to the hypothalamus. There are in fact two important pathways. One is the peripheral nervous system and the other is the rich underlying lacework of small blood vessels known as capillaries.

Signals from these two sources are integrated in the thermoregulatory section of the hypothalamus. If the reading says the body is too cold, the capillaries constrict, thereby conserving heat; if it is too warm, the capillaries dilate. Simultaneously, hormonal messages are sent to the sweat glands, ordering them to secrete moisture—sweat—through pores in the skin. When this happens, signals to the brain strongly suggest proceeding to behavioral changes, such as the donning or shedding of clothes, maintaining all the while, of course, Cushing's "inhibitions." The blood supply entering the hypothalamus provides an ongoing check of the adjustments made and, if necessary, indicates to the hypothalamus the necessary secretions for resetting temperature. Once again we can only marvel at the efficient workings of a system that has evolved over millions of years.

Constant Temperature

We share the characteristic of keeping a constant body temperature with mammals and birds, the other so-called warm-blooded animals. Of course, it's not just blood that's warm, nor is the distinction quite so clear-cut. The separation in biology is made between homeotherms and poikilotherms, from the Greek *homos,* meaning "the same," as opposed to *poikilos,* meaning "variable," with *therme* being heat or temperature. The homeotherms—birds and mammals—have high metabolisms, generate heat from within, and have elaborate cooling mechanisms to assure a constant body temperature, while poikilotherms—all other animals—do not. There are exceptions to this rule; for instance, some hot-blooded animals lower their body temperature considerably, as in the well-known case of hibernation. Nevertheless, the classification is a very good approximation, certainly good enough for us to ask why homeothermy has evolved. Requiring a more complicated brain with more sophisticated controls, it has been adopted by a very small percentage of known species. There is no single answer to why they have done it, only a set of hypotheses.

The beginnings of constant body temperature in some animals seem to coincide with the transition to living on land from an earlier aquatic existence. Life underwater is sheltered to a large extent from changes in the weather. In particular, the ambient temperature in deep water remains pretty constant. By contrast, creatures living on the Earth's surface encounter temperature variations in the course of twenty-four hours, experiencing night and day, rain and shine, wind and storm. In addition, life on the surface has evolved to the point that many animals have to rapidly make complex decisions.

Imagine an early human ancestor being chased by a lion across the savannah. In running, all the limbs have to be moved in a coordinated manner while the brain is evaluating the best strategy for survival. Run, or turn and fight with one's primitive club? How far away is that tree, and what chances do I have for climbing it before the lion reaches me? What about my family's chances of survival, and will another member of my clan come to my assistance? Should I dive into the river or do I risk being caught by a crocodile if I do that? These calculations are all going on while the limbs are moving, the body is sweating, and the lungs are straining. Both the lion and the human's thoughts and actions need to proceed in parallel and the decisions integrated into the chosen route for survival.

The master control that directs human thought and action is the brain, a fantastically elaborate circuitry of a hundred billion interconnected nerve cells; a comparable number is housed in the lion's cranium. The complex chemical reactions that activate the transmission and reception of signals are temperature-dependent, as are the various hormonal messages sent to the specialized organs. Since all the circuitry is temperature-dependent, having a constant body temperature—one with a little leeway for special circumstances—is simply the best evolutionary choice for animals as complex as we are. A fluctuating brain temperature would lead to unpredictable reactions, ones that might not occur in the same sequence if the learning had taken place at a different brain temperature. The human brain and that of other mammals and of birds—these extraordinary tools—work as well as they do because of the protected, controlled environment they are housed in. Simpler animals, with far less complex brains, have optimized their survival possibilities in other ways, but constant temperature is best for us.

Going back to my example of our ancestor trying to make decisions while being chased by a lion, he (or perhaps she) wouldn't want his arms to try climbing the tree while his legs endeavor to continue running, nor his eyes to see a rock while his nose thinks it's a lion. This is also not the time to decide it would be good to have a snack. Conversely, the lion is making exactly that decision, straining to ensure, however, that he or she is chasing an early human rather than something indigestible. Chances of survival and passing your genes on to the next generation increase as the mechanisms for multiple, simultaneous decisions and actions become flexible and rapid, yet reliable. A constant-temperature brain seems to facilitate this process.

I don't mean to suggest by this example that the elemental predator/prey interaction requires an intricate constant-temperature brain. That's clearly not true. The brains of organisms such as humans and lions operate best at constant temperature, but the reasons for such complex brains lie in the extraordinarily intricate actions, many of them social and organizational, that these creatures carry out during their lifetimes.

Nor is constant brain temperature the only reason for homeothermy. Chemical reactions generally proceed faster as temperature rises, so a higher body internal thermostat setting affords greater activity—up to a point. When excess heat cannot be shed and information is coming too quickly, the system breaks down. Over the past few million years, we, as well as other mammals and, of course, birds, seem to have found that we function most effectively in the vicinity of 100 degrees Fahrenheit.

A physiologist friend of mine told me to think about sexual behavior when faced with a puzzle about animal behavior. The hormonal reactions that control mating, procreation, and countless other commands work best at a high constant temperature in warm-blooded animals. We can even look to temperature for the answer to such fine-tuning questions as why do males have testicles in external scrotal sacs rather than in the more protected abdominal cavity. Presumably a somewhat lower temperature than 98.6 degrees is favored for sperm production.

Given that the human body works most reliably if kept at a constant temperature, why is that temperature 98.6 degrees? The rough answer is a mixture of evolutionary arguments combined with a simple understanding of how our metabolism works. Most machines are quite inefficient and mammalian bodies are no exception. Typically, more than 70 percent of the energy input into the body is converted into heat. This heat then needs to be dissipated into the environment, or else the body, like any overcharged engine, becomes overheated and stops functioning properly. We feel most comfortable in external temperatures some 20 to 30 degrees below our skin's temperature, because that differential produces a comfortable rate of heat loss; any colder and we lose heat too rapidly, any warmer and we retain too much. We correct for the former by adding clothes, blankets, and muscular activity like shivering; we correct for the latter by sweating, fanning, and, when circumstances allow, just taking it easy.

The mechanisms of heat production are intricate, another complication for the brain to oversee. While at rest, the brain and the internal organs, such as the heart, lungs, and kidneys, produce more than two-thirds of the body's heat even though they constitute well under 10 percent of body mass. In motion, the heat output by muscles can increase by a factor of ten, coming to dominate all other sources. Yet despite these dramatic changes in heat output, body temperature remains fairly constant and basic instinctive responses remain predictable. This is achieved by the body's ability to rapidly increase its transfer of heat to its surroundings while its own internal heat production is rising.

The mechanisms for heat transfer are elaborate in detail, but the basic physics principle is that heat always flows from hotter to colder objects. All objects radiate and absorb heat, more efficiently if they are dark and less if they are light: you are cooled in a large room with stone walls that are below body temperature and warmed if those same walls are above body temperature.

Conduction of heat is simply a variation of the same statement as applied to two objects touching one another; heat always flows from the hotter to the colder object. In both radiation and conduction, the rate of the flow, at least for temperature differences that are not too large, is roughly proportional to that difference—a rule that is referred to in older books as Newton's Law of Heat Flow. You lose heat to a metal bar in your hand twice as rapidly if the bar is at 78 degrees as you do if the bar is at 88 degrees because one is 20 degrees away from 98.6 while the other differs by only 10 degrees. A room with stone walls at 58 degrees feels colder than the same room with walls at 78 degrees.

It is sometimes argued that our body temperature is set at 98.6 degrees for the same reason that we feel comfortable in a room at 70 degrees. A little over two million years ago, humans emerged in Africa in sites where the median daily temperature is in the low 70s. Thus, a body temperature in the high 90s optimizes the necessary dissipation of heat generated by metabolic processes during the kind of activities hunter-gatherers pursued in those climates. You can calculate the rate at which the body produces heat during normal activities, and you can also calculate the rate at which the body transfers heat to an environment in the low 70s. Both rates depend on body temperature: a simple estimate shows the two roughly coincide when that temperature is 98.6, the point where heat in equals heat out. Later humans extended their range of thermal tolerance to cold weather by wearing furs and by the unique skill they acquired—making fires.

This adaptation to climate, however, can be at best only a small part of the reason for our 98.6-degree body temperature. Almost all other birds and mammals, with presumably very different evolutionary histories, have stabilized at more or less the same temperature. The main reasons for homeothermy are lodged in the optimization of the complicated set of chemical reactions that allow us and these other animals to carry out the complicated activities of our lives.

Into the Sahara

Efficient cooling mechanisms are just as important as warming ones in keeping a constant body temperature. The literal application of the notion that heat flows only from hotter to colder objects means that without some adjustment mechanisms, we would heat up intolerably and die if placed in an environment at 106 degrees. Instead, we survive. In fact, as we shall see later, humans survive quite well in even higher temperatures in the great Sahara desert.

Evaporation is the key, carrying away some 25 percent of our metabolically generated heat even in coldish surroundings, and more as the temperature rises. To understand how evaporation works, think of our body fluids as pools of liquid—molecules of water moving about and colliding with one another. It turns out that the temperature of the water can be related to the average energy of motion (the so-called kinetic energy) of those molecules. If some of the faster water molecules escape, the average energy of the remaining liquid, which then reequilibrates by conduction, is lowered and the overall temperature of the fluid drops. This is why evaporation cools, but the process only works if the air above the fluid is sufficiently dry. If fast molecules of water vapor reenter the pool of liquid as quickly as they leave, the cooling effect is erased.

The process of evaporation is so important because of the large amount of heat absorbed in changing water into steam. Calories are units of heat. It takes one calorie to raise one gram of water 1 degree Celsius, or 100 calories to go from 0 to 100 degrees Celsius. But more than 500 calories are necessary to change a gram of water at 100 degrees Celsius into a gram of steam at the same temperature. In other words, the amount of heat employed in the simple transition from water to water vapor is more than five times as great as that in going all the way from freezing to boiling temperature. This means the conversion of water to water vapor in the body is a very efficient way to cool down.

Not surprisingly, animals that have to lower their temperature use this basic mechanism, adapting it to their own needs in an almost infinite number of ingenious ways. One way to help the process of cooling by evaporation is by fanning, or blowing the faster water molecules away as soon as they leave the surface and before they reenter the liquid. This is the same process we follow in blowing on the surface of a coffee cup or a too-hot bowl of soup.

Of course, humans made an adornment out of what began as a necessity: form follows function and decoration follows form. From the Latin *vannus,* a fan is our oldest known cooling instrument. A bas relief in the British Museum shows Sennacherib being fanned with great feathers, long since turned to dust. But the handles—from almost two millennia B.C.E.—have been preserved. The folding fan, so elegantly opened and closed with a snap of the wrist, seems to have first appeared in Japan and then in China, where it was an homage to have a distinguished guest write a short message on your fan to mark a special occasion. Queen Elizabeth I of England had her portrait painted holding a large feather fan. By the eighteenth century, fans were so *à la mode* that Europe's major cities had painters who only decorated fans. Fan handles were carved out of ivory and mother-of-pearl, with jewel adornments and lenses mounted on the end. At formal dances an elaborate etiquette of signaling through positions of the fan was established, a kind of semaphore code. But first and foremost, the fan was and still remains the simplest instrument for cooling.

Bees also employ this fanning strategy. The temperature inside their nests is carefully regulated. In the summer they cool the hive by fanning their wings to stimulate circulation, but when the temperature reaches the 80s, fanning isn't enough. Then bees leave the hive in search of water. On return they regurgitate the water, forming thin films and droplets that they then fan, driving the moist air out of the nest. E. O. Wilson cites an experiment in which bees, given unlimited access to water, were able to maintain a hive temperature of 85 degrees although the outside temperature reached 160 degrees. It's not heat that keeps bees out of the desert: it's the lack of water.

Fanning accelerates cooling, but the first step obviously has to be production of fluid on a surface from which it can evaporate. A few kangaroos and some rats cool themselves by licking their fur, allowing the saliva to evaporate, but the two most widely used evaporation techniques are panting and sweating.

Birds lack sweat glands and certain mammals, such as dogs, have very few; they rely on short, rapid, and shallow breathing called panting to stimulate evaporation from their throats. This technique has a few advantages, one of which is far from obvious: it literally aids in keeping a cool head. The small East African gazelle running in the plains at full speed for five minutes generates so much internal heat that its core temperature rises from 102 degrees to slightly over 110 degrees. Nevertheless, its brain, fed by arterial blood coming from the body at 110 degrees, remains more than 5 degrees cooler. The ingenious brain-cooling adaptation is a by-product of the rapid breathing of the animal in flight. The main blood conduit from the core to the brain is the carotid artery, which branches into hundreds of small arteries at the base of the skull before consolidating into one again as it enters the brain. In the passageway in which it branches out, the blood is cooled by the rapid flow of air in the adjacent throat. This interesting cooling mechanism is well suited to maintaining optimal functioning of the fleeing animal's decision-making ability. The gazelle keeps an approximately constant brain temperature even while the rest of its body is heating up. Thus, the animal's first efforts are directed to keeping a constant temperature in the control center, the brain. The rest of the body has a little more leeway.

Panting has another advantage over sweating. The fluid secreted in sweat carries away precious salts; hence the usual admonitions to drink fluids containing appropriate minerals when sweating profusely. In panting, on the other hand, the body minerals in the saliva stay in the body. However, panting has its own disadvantages, one of which is that it requires muscular activity, an activity that itself generates heat. (This is alleviated in part by the rapid shallow breathing.) No solution is perfect. All are adaptations that animals have evolved over long periods to optimize their chances for survival.

Most large mammals sweat, some more visibly than others. Even camels do, though it isn't very noticeable in the dry desert air because the water vapor disappears almost immediately. Humans have lost their fur almost entirely, leaving a naked skin. This outer covering has some two million sweat glands distributed around the body, with higher density in the palms of the hands and lower density in other areas. Under the control of the hypothalamus, the glands secrete a slightly salty fluid. That secretion is not voluntary nor is it stimulated solely by environment; stress or nervousness also induces sweating. Regardless, it is a very efficient cooling mechanism when increased metabolic activity generates body heat that has to be dissipated quickly. Increased sweating may not be desirable when you're in a clean shirt and slacks en route to an office, but rapid cooling to aid flight may have been very useful for our ancestors surviving in the wild.

There is a caveat about the effectiveness of cooling by sweating, one we have already met: more fast molecules have to be leaving the body surface than arriving. If the outside air is too humid, sweat rolls off the body without any of the needed evaporative cooling.

Dehydration can also pose a problem. We produce, on the average, about a quart of sweat per day, even if we are unaware of it, though this figure can fall to almost zero and rise to as much as four gallons depending on weather and level of activity. Loss of an amount close to that maximum carries with it the danger of serious dehydration, possibly requiring intravenous replenishment of fluids.

In one of the regular "Commentary" articles that Philip Morrison used to write for *Scientific American,* he illustrates the power of sweating and fanning by considering those ultimate athletes, the Tour de France bicycle racers. Morrison tells how the great Eddy Merckx, five-time winner of the Tour, fared in a lab experiment in which he rode only a stationary bike. The man who could ride up and down hills day after day for six hours at a time collapsed in a pile of sweat after a single hour in a breezeless indoor gym. Why? Morrison does the numbers.

Racers eat the equivalent of eight square meals a day to get needed energy, since bicycle racing uses up about one thousand Calories an hour, ten times the amount burned in an hour sitting at a desk. (Note, incidentally, that Calorie spelled with a capital C is the conventional, sometimes confusing, way of denoting a kilocalorie, or a thousand calories. A Calorie is the amount of heat needed to raise a kilogram of water through 1 degree Celsius. Technically, the water should be at

15 degrees, but that's a fine point. Even more technically, there has been a recent redefinition of the Calorie in terms of equivalent mechanical work.)

During the twenty-two-day Tour de France race, the cyclists neither gain nor lose weight, so what happens to the energy? Only 25 percent of it goes into the mechanical work of overcoming air drag and propelling the bicycle and racer forward. A full 75 percent is dissipated as extra body heat—so much heat that the racer needs ten quarts of water to evaporate from his skin each day of the race in order to stay at constant temperature. This requires continual drinking, but the racer also needs a strong breeze to help the evaporation; speeding along at twenty-five miles an hour or more provides this breeze. No breeze means saturated vapor pressure, no evaporation, and heat buildup. The result is that the Eddy Merckx who can ride at top speed for eight hours falls off a stationary bike in a state of total exhaustion after sixty minutes. Today, survivors of athletic club spinning classes can attest to the same effect.

Evaporative cooling is also the key to survival in the great Sahara desert, where temperatures rise above 130 degrees Fahrenheit. No need to worry about saturation of water vapor pressure in the very dry desert air. A person will lose two gallons of fluid per day simply sitting in the shade of a Saharan date palm gently fanning him- or herself. As much as four gallons evaporate in a day of moderate exercise. Extreme exercise is out of the question. And this water supply needs to be continually replenished: the first symptoms of dehydration set in after loss of a pint, fatigue and fever begin after loss of a gallon. After a two-gallon loss, dizziness and difficulty in breathing start, and by three gallons, the point of no return is reached unless prompt medical treatment and intravenous rehydration is available. Without replenishment of water, one summer day of walking in the Sahara is fatal.

Nevertheless, people thrive in the desert. Part of the adaptation to hot, dry climates can be achieved by almost anyone over the course of five to ten days through gradual exposure, for a few hours a day, to high temperatures. In essence, one trains the body to sweat more. As physiologists and experts in animal temperature Carl Gisolfi and Francisco Mora put it, "This is perhaps the most remarkable physiological adjustment that humans are capable of making and is attributable in large part to the evolution of the sweat gland."

Not only is the level of sweating increased in this adaptation, but also the sweat itself becomes less salty. This serves the double function of preserving the sodium, potassium, and other minerals in sweat that the body needs for proper functioning, and of increasing thirst, in turn stimulating greater drinking. More drinking of water means more evaporation and a cooler body.

Desert-inhabiting humans are also aided in their fight for survival by a unique animal that can drink almost thirty gallons of water in ten minutes and then distribute it through its body. That animal is, of course, the camel, a remarkable example of adjustment to desert life. It stores water and then hoards it: it can live for two weeks between drinking sessions, surviving by calling on its internal reservoir. The camel is also helped by a series of water-conserving strategies; for example, its urine and feces have very little fluid in them, and it generally keeps its mouth shut and narrows its nostrils to slits. However, the most extraordinary adaptation of this remarkable animal is its ability to adapt its thermostat to internal fluid levels. A camel with a full supply of water will keep its body temperature steadily between about 97 and 100 degrees Fahrenheit, relying on evaporation for cooling. If its water supply is low, it will shift its acceptable temperature range, going from as low as 93 degrees during the night to 106 degrees during the hottest part of the day. Tolerating a considerably higher body temperature while in the sun diminishes the need for evaporative cooling. The heat accumulated during the day is then offloaded as much as possible at night. Here is an example of an animal that prefers to keep body temperature constant. If stressed, however, it adapts to the new conditions.

Smart desert travelers found that evaporation also kept their water supplies cool. The wonders of evaporation, even in a desert climate, were not lost on Benjamin Franklin, a man who seems to have been curious about how just about everything works, from governments to lightning rods. While in England defending the rights of the American colonists, Franklin, who incidentally also founded the university at which I teach, performed experiments in which he wet a thermometer with ether. He then blew on it with a bellows until a thin layer of ice began to form on the ball of the thermometer. In a June 17, 1758, letter to a friend, John Lining, Franklin describes some of his experiments and reflects on what desert travelers had already learned:

> From this experiment one may see the possibility of a man freezing to death on a warm summer's day, if he were to stand in a passage through which the wind blew briskly, and to be wet frequently with ether, a spirit that is more inflammable than brandy, or common spirits of wine. It is but within these few years that the European philosophers seem to have known this power of nature, of cooling bodies by evaporation. But in the east they have long been acquainted with it. A friend tells me that there is a passage in Bernier's *Travels through Indostan,* written near one hundred years

> ago, that mentions it as a practice (in travelling over dry desarts in that hot climate) to carry water in flasks wrapt in wet woolen cloths and hung on the shady side of the camel.

Franklin then goes on to muse about his ability to withstand summer days in Philadelphia when the temperature reached 100-plus degrees. (I have experienced many of those Philadelphia days myself, so I was particularly interested in his views on this subject.) He concludes that the cooling power of evaporation must make the difference or, as he put it:

> And I suppose a dead body would have acquired the temperature of the air, though a living one, by continual sweating, and by the evaporation of that sweat, was kept cold.
>
> May not this be the reason why our reapers in Pennsylvania, working in the open field in the clear hot sunshine common in our harvest-time, find themselves well able to go through that labour, without being much incommoded by the heat, while they continue to sweat, and while they supply matter for keeping up that sweat, by drinking frequently of a thin evaporable liquor, water mixed with rum; but, if the sweat stops, they drop, and sometimes die suddenly, if a sweating is not again brought on.

I should add that Benjamin Franklin, ever practical, was also interested in how to stay warm when it's cold outside. The Franklin stove attests to this concern.

Into the Antarctic

As with cooling, humans have multiple adaptation mechanisms to create and preserve warmth. While shivering generates additional internal heat, our external cover helps by trying to seal off the loss of warmth to the outside. When stressed by cold surroundings, the blood flow to the body surface is rapidly constricted and skin temperature drops. This reduces heat transfer to the outside since the temperature difference between air and skin is lowered. In a sense, the outside of our body tries to form an insulating cover for the vital internal organs. These are short-term defenses against the cold.

There are also long-term human adaptations to cold weather, such as adding a little extra fat in the winter, though they are nothing like the dramatic ability to hibernate that some mammals and birds possess. In that state, an animal lowers its body temperature by at least 20 degrees and goes without food or water for months. Bears can hibernate for six months. During this whole period the animal retains its ability to return at will, without external heat sources, to normal body temperature and to resume normal metabolism.

Many studies illustrate human long-term metabolic adaptation to hot weather, but comparatively few show how the human body adjusts to uncomfortably cold surroundings. The best-studied case I know of is that of the Japanese and Korean women who earn their living by year-round diving into deep ocean waters to harvest plant and animal life. Known simply as *ama,* these women begin their careers in their early teens and continue diving until their sixties. They now wear wet suits, but until the late 1960s, they dove into 50-degree water wearing simple cotton suits. Studies by Suki Hong in the 1960s showed that these women elevated their metabolic rates by 30 percent during the winter months. The purpose was presumably to generate additional internal heat in order to balance the losses incurred in dives. The argument that this was cold acclimatization is confirmed by the fact that the *ama* ceased elevating their metabolic rates once they started wearing wet suits.

Swimming or diving is a particularly interesting case because the evaporation factor is removed from the calculations of warming and cooling. In 1987, the great long-distance swimmer Lynne Cox decided to swim the 2.4-mile gap between two islands in the Bering Strait—one part of Alaska and the other in Siberia. In doing so, she would cross the International Date Line and symbolically link the United States to the Soviet Union. The strong currents meant the effective distance might be more like five miles, which was still no problem for a strong swimmer like Cox. But the water temperature was another matter. It was 44 degrees at the surface, but the churning of the deeper water meant that it could be as low as 34 degrees in spots. This was long-distance swimming in ice water.

Before embarking, Cox took the precaution of swallowing a thermosensitive capsule equipped with a transmitter to be monitored by a doctor in an accompanying boat, to ensure Cox didn't develop hypothermia, a potentially fatal condition that follows when body temperature drops below 93 degrees. She completed the swim in a little over two hours, maintaining normal body temperature throughout that period. Mikhail Gorbachev cited her bravery in a speech he made at a White House banquet later that year: "She proved by her courage how close to each other our peoples live."

How did Cox do it? Part of her success was due to a physique ideally designed for such a swim. At five feet six and 180 pounds, she had a percentage of body fat almost double that of the average woman. In addition, the fat layer was evenly distributed, providing a natural thermally insulating layer between her interior organs and the cold outside. This explains how she survived, although surely the human spirit was a big ingredient as well.

Cox's insulating layer, remarkable as it was, pales in comparison to a harp seal's. These magnificent swimmers keep their body temperature at 98 degrees and their metabolism unchanged even in freezing Arctic waters. What protects them is the quarter-inch-thick layer of blubber immediately below the skin. Once past that layer, the seal's body temperature is almost identical to its core temperature. In other words, the harp seal's skin temperature is essentially the same as that of the surrounding water, but the quarter-inch insulation manages to keep the interior as much as 70 degrees warmer.

Lest you think of the seal's layer of blubber simply as a surfer's wet suit, ponder how the seal manages to survive in warm waters without any basic changes in metabolism. The secret? The blubber is permeated by a system of capillary blood vessels that close while the seal swims in cold water and open while it moves in warm water, or while it exercises vigorously or sunbathes on a rock. No impermeable wet suit, the seal's blubber is an active and extraordinarily efficient thermal regulator. Although most adaptation to cold in humans seems to proceed by elevation of internal metabolism, as shown by Hong's study of the *ama,* I wouldn't be surprised to learn that swimmers training in cold waters to cross the English Channel develop a slightly greater ability to constrict skin capillaries. But, we are nevertheless very different from seals.

According to the physiologist Knut Schmidt-Nielsen, the difference in cooling mechanisms between aquatic and terrestrial animals lies in the relative locations of the insulator and the skin, the organ through which heat is dissipated to the outside. Seals have their insulator, the blubber, inside their skin. Terrestrial animals have their insulator, fur, on the outside. We humans have adapted a flexible middle road, though we clearly lean to the terrestrial.

Extreme temperature adventures such as that of Lynne Cox highlight the mechanisms we use for heating and cooling. We all have our favorite stories of extremes, but tales of survival in ice and snow—the experiences of South Pole explorers and climbers in the world's high mountain ranges—are the ones that move me most. The first step in staying warm in arctic weather is good clothing. Clothes are, of course, not a simple matter; the evolution of clothing and protection has made tremendous progress with the development of lightweight synthetic fibers, a point driven home to me every time I look at the wooden ice ax, the canvas tent, and the woolen clothing my mountaineer father-in-law used on a 1930 Himalayan expedition. The key to dressing well in very cold climates is remembering that air conducts heat poorly. You can see that by the insulating efficacy of double-paned windows. But air from movement or winds, unless neutralized, continuously replaces the warm layer of air surrounding your body. Many layers of clothing or a fluffed up, preferably dry, down jacket keep a warm thermal layer undisturbed.

For sheer grit, few stories compare to the Antarctic tale of Apsley Cherry-Garrard, chronicling his experiences as a member of Robert Scott's ill-fated 1910 expedition. The chapter called "The Winter Journey" describes a six-week Antarctic winter trip Cherry-Garrard and two companions took across the ice to reach the rookery of the large emperor penguins. They believed these penguins provided a significant evolutionary link between birds and reptiles. They also thought it would be particularly important to examine the penguin eggs to ascertain their embryological development. The added challenge was that the nest is the penguin itself. These rare birds lay their eggs, cradle them on their feet for safety, and press down on them with their breasts to provide cover and warmth—all in mid-Antarctic winter.

The trio set off from the expedition's winter refuge to find the penguins, dragging two sleds with 750 pounds of equipment (the weather was too rough for dogs). Cherry-Garrard was so nearsighted that he was functionally blind without his glasses, but he couldn't wear them because they would frost up. However, this posed no problem because there is no sunlight in the Antarctic during the month of July.

The temperature regularly dropped to –70 degrees Fahrenheit and once reached 77 below. All three men were frostbitten and had large blisters whose fluid turned to ice, but they kept on going. As Cherry-Garrard says:

> The trouble is sweat and breath. I never knew before how much of the body's waste comes out through the pores of the skin. On the most bitter days, when we had to camp before we had done a four-hour march to nurse back our frozen feet, it seemed we must be sweating. And all this sweat, instead of passing away through the porous wool of our clothing and gradually drying us off, froze and accumulated. It passed just away from our flesh and then became ice.

Cherry-Garrard realizes this at the beginning of the trip. He later describes emerging from his tent to pull the sled:

> Once outside, I raised my head to look around and found I could not move it back. My clothing had frozen hard as I stood—perhaps fifteen seconds. For four hours I had to pull with my head stuck up and from that time we took care to bend down into a pulling position before being frozen in.

Sleeping bags froze and stayed frozen, coffins of ice in which it was almost impossible to sleep. Despite all this they completed the journey and even brought back three emperor penguin eggs. The survivors of the Scott expedition returned to England just in time for World War I, but as one member of the Antarctic team put it, "The trenches at Ypres were a comparative picnic." Incidentally, this story is a reminder that sweating can occur at any temperature.

While reading the story, I began to think a little bit more about the emperor penguins, wondering how they manage to survive, sitting on the ice hatching their young. The answer turns out to have a few interesting twists of its own. These large birds, weighing as much as eighty pounds, are the most cold-resistant birds known. In addition, though they feed only at sea, the rookeries in which they lay their eggs can be fifty or more miles away from open water, requiring a long march by not very graceful walkers. Though of course the female lays the egg, she returns almost immediately to the sea to replenish her food supply, coming back to the rookery when the egg hatches. Before leaving, she places the egg on the feet of the male who nestles on top of it, guarding and warming his heir for as long as two months. During that period of immobility, the male eats no food and has no obvious protection from Antarctic winter conditions.

Emperor penguins huddling for warmth on the Antarctic ice

How does the male manage the fast and how does the chick survive the brutal cold once it emerges from the egg? The male makes no dramatic changes in metabolism; during his fast he burns up more than a third of his body weight, the fuel necessary for survival. A quick calculation shows, however, that he would need almost twice as much fat to survive two immobile months in the rookery, so that can't be the whole answer. The missing ingredient is very simple: the penguins huddle together, sheltering each other against the cold and the wind. Once hatched, the babies also huddle together until they are ready for the trip to the sea. The penguins have found a simple, elegant solution, one found by countless other species of all sizes and shapes.

Even though the penguin's shelter hasn't changed much, Antarctica's human refuges are now a far cry from those of Cherry-Garrard's days. The South Pole Station has a sauna and a Three Hundred Degree Club, with exclusive membership. To join, you have to expose your body to a 300-degree temperature change in under a minute. You do this by sitting in a plus-205-degree sauna and then sprinting outside naked to minus-100-degree weather, listening to your skin crackle. You pause and then rush back in.

No other animal species lives everywhere from the poles to the equator nor is as resourceful as humans are. Nevertheless, our control mechanisms do sometimes break down. One of the most common occurrences when that happens is fever.

When Things Go Wrong

I recall walking with my mother in late fall in Florence, Italy, where we lived at the time. I was nine years old and, though warmly dressed, I was shivering. When I told my mother I wasn't feeling well, she put a hand to my forehead and said that I had a fever and that we needed to go home immediately. Then she added, as we passed a bookstore on the way to the trolley, that she thought I was old enough to buy a real book for myself, my first. It was Jules Verne's *Mysterious Island;* the next three days of fever were wonderful, and a lifelong love of books had begun. In the book, the island's colonists even discover "fever-trees," trees that supposedly prevent fevers, though they don't cure them. My recovery was perhaps prolonged by my wanting to finish the book; the cause of my fever was not ascertained, but as usually happens, the fever ended anyway.

The truth of the matter is, even the wizardry of modern medicine is often unable to find a fever's cause. Doctors say they often find patients with fevers whose source cannot be found. The criterion for a mysterious fever is that it lasts at least three weeks at a level of at least 101 degrees with no explanation, despite one or more weeks of investigation in the hospital. These mysterious fevers are particularly troublesome, but even if the cause of a disease is known, it's important to record the course of the fever.

Temperature, blood pressure, pulse, and respiratory rate are the four functions recorded on the medical charts at the foot of all patients' hospital beds because they often reflect the course of a disease. Even if we can't identify the disease and don't know how to treat it, these four let us know its progress and often give the first hints of recovery.

High temperatures pose particularly serious dangers. In acute infections like meningitis, typhoid, or pneumonia, an unchecked fever can rise to 107 degrees or higher, posing a serious risk to the patient independent of the disease that has caused the fever. Tremors, delirium, and convulsions can set in. Treatment needs to be swift but also, as *Harrison's Principles of Internal Medicine* emphasizes:

> It is in the diagnosis of a febrile illness that the science and art of medicine come together. In no other clinical situation is a meticulous history more important. Painstaking attention must be paid to the chronology of symptoms, to the

use of drugs (including those taken without a physician's supervision) or treatments.

This is true because an uninformed treatment of the fever may do more harm than good.

Fever-causing substances are known as pyrogens, sharing with pyrexia, pyrotechnics, and pyromania the Greek root for fire, pyro. A further distinction is made between *exogenous* pyrogens, produced outside the body, and *endogenous* ones, which come from within the body. Pyrogens, which may be bacteria, stimulate fever by evoking the release in the host of a class of chemicals known as cytokines, which are then carried through the bloodstream. Eventually they reach the hypothalamus. There, in Cushing's "mainspring of primitive existence," they produce yet another chemical, prostaglandin. This resets the body's thermostat to a higher temperature: it induces fever.

In resetting body temperature, opposite reactions to fever and exercise occur. In both cases the core body temperature goes up, but with exercise the body sweats, an attempt to lower that temperature to the normal set value. With a fever, it shivers. The body is now trying to generate internal heat by involuntary muscle contractions, in an effort to bring core body temperature into line with the set value; in other words, the body "thinks" it is supposed to be hotter.

There are two ways to change the set temperature value in the hypothalamus back to normal when you have a fever. One is to remove the pyrogens—that is, kill the bacteria that are leading to the cytokines. The second is to administer aspirin or similar drugs that inhibit the synthesis of prostaglandin. In plain language, destroy the message or kill the messenger.

Though aspirin as such has only been known for a little over a century, related products were recognized very long ago. Hippocrates is generally regarded as the father of modern medicine; hence the Hippocratic oath embodying the code of medical ethics. He treated fevers and pains in the fifth century B.C.E. by administering willow bark extracts. The Latin name for willow is *salix* and the active ingredient in the bark is salicin. Unfortunately, salicin causes stomach upset, a situation remedied in 1897 when a Bayer Company chemist synthesized acetylsalicylic acid, a much improved version of salicin. The new product, known as acetylspirin, was marketed as aspirin by Bayer in 1899. First sold without a prescription in 1915, it had far fewer noxious side effects than salicin. Its value was very quickly realized, although its role in inhibiting prostaglandin wasn't recognized until the work of John Vane in the 1970s. He received the 1982 Nobel Prize in medicine for his insights.

It's worth emphasizing once again, however, that even with all the tools of modern medicine, a clear diagnosis of the cause of fever often escapes both the patient and the doctor. Sometimes, fortunately, treatment is swift and improvement is dramatic. Until the twentieth century, however, there was very little in the way of effective drugs to treat infection. The use of antibiotics, now complicated by the emergence of resistant strains of bacteria, is little more than fifty years old. I can't think of a better way of emphasizing how recent this is than by quoting a June 9, 1999, *New York Times* obituary:

> Anne Sheafe Miller, who made medical history as the first patient ever saved by penicillin, died on May 27 in Salisbury, Conn. She was 90.
>
> In March 1942, Mrs. Miller was near death at New Haven Hospital suffering from a streptococcal infection, a

> common cause of death then. She had been hospitalized for a month, often delirious, with her temperature spiking to nearly 107 while doctors tried everything available, including sulfa drugs, blood transfusions and surgery. All failed.

The story then goes on to recount how, though Sir Alexander Fleming had discovered penicillin in 1928, its therapeutic powers were not fully appreciated until the apparently miraculous recovery of Mrs. Miller after its use. Fortunately for her, the cause of her fever was known and a treatment was available.

Although Hippocrates was the first recorded doctor to treat fevers, Galen was the most influential figure in the early history of the subject. Born of Greek parents in Asia Minor during the reign of the Roman emperor Hadrian, Galen was a great expositor, codifier, and teacher. He also wrote voluminously, summarizing his observations in books such as the lengthy *On the Usefulness of the Parts of the Body.* Until the seventeenth century, these tomes dominated thinking about the human body every bit as much as Aristotle's writings did philosophy or Ptolemy's astronomy.

The Greeks thought of earth, air, water, and fire as the constituents of matter, and hot, cold, dry, and wet as the basic sensory experiences. Taking off from there, Galen formulated a doctrine of four basic body humors: blood, yellow bile, black bile, and phlegm. Each humor, which one should imagine as a kind of body fluid or sap, led to a characteristic physiognomy, behavior, and even coloring. Blood was associated with air, morning, and spring while yellow bile was tied to fire, midday, and summer. Black bile corresponded to earth, evening, and fall while phlegm was joined to water, night, and winter. Personality or temperament was also viewed as having four major types: sanguine, choleric, melancholic, and phlegmatic, associated respectively with blood, yellow bile, black bile, and phlegm. To give an example, Albrecht Dürer's 1526 panels of *The Four Saints* are probably intended as illustrations of the four temperaments, with St. John as sanguine, St. Mark choleric, St. Paul melancholic, and St. Peter phlegmatic.

Given yellow bile's association with fire, it's not surprising that Galen thought an excess of yellow bile caused fever. Humoral medicine, as he practiced it, strove for equilibrium to be restored. Once disturbed, balance was to be restored through the application of heat or cold, wet or dry, feeding or purging and bleeding. Humoral medicine is presently undergoing a rebirth of sorts, sometimes in association with modern medicine and other times as a derivative of related practices such as the Indian Ayurvedic medicine, practiced hundreds of years before Galen. Though the idea of yellow and black bile should not be taken literally, and fevers are certainly not induced by an excess of yellow bile, the notion of equilibrium of one's systems may have a good deal to recommend it.

Virulent microorganisms that enter the body can and often do destroy that equilibrium. Until doctors understood the activity of germs, surgery often did more harm than good, mainly because infections followed operations performed in nonsterile environments. We should remember that the germ theory of disease, largely the work of Louis Pasteur, is little more than a hundred years old.

The distinguished microbiologist René Dubos, in his biography of the great man, quotes the statistic that ten thousand of the thirteen thousand amputations performed on French soldiers in the Franco-Prussian War proved fatal. Pasteur himself, on visiting hospital wards, came to realize that transmission of germs by contamination of the hands of the surgeon or dressings was even more important than airborne transmission. In a famous lecture to the French Academy of Medicine in 1878, Pasteur said,

> If I had the honor of being a surgeon, impressed as I am with the dangers to which the patient is exposed by the microbes present over the surface of all objects, particularly in hospitals, not only would I use none but perfectly clean instruments, but after having cleansed my hands with the greatest care, and subjected them to rapid flaming, which would expose them to no more inconvenience than that felt by a smoker who passes a glowing coal from one hand to another, I would use only lint, bandages and sponges previously exposed to a temperature of 265 to 300 degrees.

Pasteur had realized the dangers of infection, but it was a young Scottish surgeon, Joseph Lister, who is generally given credit for developing and systematizing the notion of antiseptic surgery to curb the infections that followed wounds or surgery. He did acknowledge freely his debt: to quote from an 1874 letter from Lister to Pasteur, "Allow me to take this opportunity to tender you my most cordial thanks for having, by your brilliant researches, demonstrated to me the truth of the germ theory of putrefaction, and thus furnished me with the principle upon which alone the antiseptic system can be carried out."

Though we often think of bacteria as invaders of our bodies, many strains live tranquilly within us, causing no problem. Even the notorious *Escherichia coli,* commonly known as *E. coli,* is ubiquitous. It happily takes up residence in our colons: it's hard to find a mammal not blessed with these visitors. If *E. coli* finds its way outside the gut, moving, for instance, to the urinary tract, it can cause infections, but the problem is usually not too serious.

Some strains of *E. coli* are noxious, and occasionally a really bad strain develops. *E. coli* O157:H7 is particularly virulent. It was first identified in the early 1980s when some people became violently ill after eating contaminated hamburgers. *E. coli* in the gut of cattle doesn't affect the animal's health, but slaughtering is a messy business, and it's easy to see how some contamination can slip in, particularly in cuts of meat that are processed by grinding. After the first recorded appearance of *E. coli* O157:H7 and the panic that followed, treatment was sought as well as a means of preventing contamination. Fortunately, there's an easy solution for hamburgers: cooking for fifteen seconds at 160 degrees Fahrenheit is enough to kill the bacteria. But there are still 75,000 cases a year of O157:H7 infection in the United States.

Heating to 160 degrees is not always practical—witness the example of surgeons trying to flame their hands. That's one place where the medicine of the future may help. After Galen and Pasteur, we now have genomic sequencing. The January 25, 2001, issue of *Nature* has an article providing the full genome sequence of *E. coli* O157:H7. As the authors say, "The severity of the disease, the lack of effective treatment and the potential for large-scale outbreaks from contaminated food supplies have propelled intensive research on the pathogenesis and detection of *E. coli* O157:H7." The hope is that the full sequencing will lead to better diagnostic tools and even to treatments. In the meantime, be sure to cook those hamburgers.

Beyond heating, we don't have a treatment for *E. coli* O157:H7. We don't have treatments for many causes of fever either, but much more surprising is the fact that we don't even really understand why we get fevers. The explanation for a fever obviously isn't just the simple one that increasing the body's temperature by a few degrees kills the invading bacteria. Some bacteria, such as pneumococci, are particularly sensitive, growing poorly above 106 degrees, but by and large, temperatures have to be raised well above the limits of human tolerance to kill bacteria. Sterilization of medical and dental instruments or the cooking of meats and fowl typically involves temperatures ranging upward of 160 degrees (remember Pasteur's advice to surgeons). So if killing the bacteria isn't the reason for temperature elevation, what is?

Shocks from Heat

Excess or prolonged fever is harmful to a patient, but it is less clear that a fever is always harmful. If that were the case, why has it survived through millions of years of evolution? The costs to our bodies seem to outweigh the benefits: every increase of a degree in temperature means approximately a 7 percent greater oxygen demand, more need for fluids, and more stress on the heart and other organs. Fever diminishes mental functioning and can produce delirium in even nonfatal episodes. It shocks the system, so much that mental patients were sometimes treated by having fevers induced. The 1927 Nobel Prize in medicine was awarded to Julius Wagner-Jauregg for his observation of psychiatric improvement after infection, a study that led to the treatment in the 1930s of late-stage neurosyphilis patients with malaria. As one doctor remembers the care of mental patients early on in his career,

> We treated all patients with the tools that were available. Colonic irrigation was still used. So was fever therapy. We had a strain of malaria that we would inoculate patients with. Later on we used a typhoid strain. We'd inject a typhoid vaccine and within hours patients would experience

> nausea, vomiting, diarrhea and fevers of 104 to 105. We'd do that for eight or ten weeks, two or three times a day. We did it to take the starch out of disturbed patients.

In other words, the treatment may have been intended more as sedation than anything else.

But there is some evidence that fever enhances the functioning of the immune system; white blood cells, the system's agents, move more rapidly as temperature approaches 104 degrees, but that's only one of the possible reasons for the evolution of the fever response. P. A. Mackowiak has suggested that fever sometimes plays a protective role: a mild infection heals rapidly with perhaps a slight enhancement of the immune system, but a raging high fever that leads to a rapid death of the afflicted individual helps limit the spread to the individual's kin of a violent contagious infection. If the explanations for fever are this general, we should expect that the fever response holds for a wide variety of species in the animal kingdom. Let's look there for clues.

Mammals and birds, the warm-blooded animals, develop fevers, but cold-blooded animals also react to infection. Unable to raise their core temperature by internal mechanisms, lizards simply move to warmer locations when injected with a harmful bacterium. Fish also display this behavior; if they are prevented from making these beneficial transitions, mortality rates from infection go up significantly. Even insects manifest this behavior: studies of subjects ranging from the Madagascar cockroach to the American migrating grasshopper show the same movement toward a warmer environment after infection has set in.

The generality of thermal sensitivity and of the capacity for thermal regulation is not surprising since the origins of the hypothalamus, aside from other parts of the brain, go very far back in time. How far back? Perhaps 550 million years, to the beginnings of vertebrates, those creatures with skeletons, spinal cords, and skulls housing a brain.

Vertebrates' closest living relative from that earlier era seems to be a two-inch-long silvery organism that lives in a hole dug in the sand. Going by the scientific name of *amphioxus,* this lancelet doesn't have a brain or a skeleton, but it does have a nerve cord supported by a rod of harder tissue stretched along its back. The cord has a swollen tip, perhaps an early version of a brain. Nicholas and Linda Holland have been gathering these lancelets in Florida waters and studying that tip with the techniques of modern molecular biology. They find that the same genes that subdivide the vertebrate brain into forebrain, midbrain, and hindbrain determine cell placement and the overall organization of the tip. It seems that nature solved a problem a long time ago and is using that solution over and over again.

The similarities of lancelets to vertebrates are enough to place them together in the chordate phylum. Lancelets also have a brain, but does it work like ours? The genetics research of the Hollands is proceeding in parallel with a detailed analysis of the lancelet brain by neuroanatomist Thurston Lacalli. He has been slicing the animal's cord tip into fine sections and studying how the neurons are wired together. It's painstakingly slow work. As Linda Holland says, "It's like taking a 747 and chopping it up a millimeter at a time," but the fine work pays off. Lacalli claims the neuronal structure matches that of vertebrates. There's a cluster of cells near the tip that form a kind of primitive eye, not good enough to see with, but probably able to distinguish shadows. There's even something that acts like a primitive hypothalamus, issuing simple orders like swim or eat. Maybe there's even temperature sensitivity.

The communality of the anatomy of brain structure explains a great deal about the shared response of many species to a rise in core temperature, but the effort to move toward a warmer location as a response to infection seems to be so universal that it becomes attractive to look for its origins beyond the vertebrates. After all, insects also respond to infection by seeking warmth. If the mechanisms are this general, a good place to find a possible explanation is the fruit fly. Drosophila, the workhorse of twentieth-century genetics, reproduce in a few days, a conveniently short time, and thrive on a diet of rotten bananas.

In the 1930s, a discovery that may eventually lead to a much deeper understanding of temperature was made in drosophila. All cells in a fruit fly carry four pairs of chromosomes, but the cells in the fruit fly's salivary glands are special. In those little throat protuberances, thousands of copies of each chromosome are precisely aligned, creating a giant repeated replica of the original. Without a magnifying glass you probably can't see specks of color on a foot-long piece of string, but when you take five thousand identical copies of the string and put them side by side, the speck becomes an easily visible line of color across all pieces. Similarly, the features on ordinary fly chromosomes are too small to be seen by a light microscope, but the giant *polytene chromosomes* in the salivary glands show the fine structure in all its details.

In 1962, F. M. Ritossa noticed that the fruit fly's polytene chromosomes puff up when the fly is exposed to temperatures somewhat higher than those at which it normally thrives. The puffing continues for half an hour, as the chromosomes swell to twice their original size, and then abates. That's the classical genetics side of the picture.

On the molecular biology side, Alfred Tissieres and Herschel Mitchell showed in 1974 that the puffing was accompanied by the abundant production of new kinds of proteins. These came to be called "heat shock proteins," or, simply, *hsp*. The role they played was at first unclear, but an interesting picture began to emerge in the years that followed. To do the cell's work, many different proteins must move around and interact with each other in just the right way. This requires that the DNA code dictate the right chemical formulas. But that's not enough; the proteins also need the right folding over of the structure-determining chains of amino acids. Without the correct spatial configuration, the molecules don't fit together and can't even properly recognize one another.

The ways in which long protein molecules fold up are still a mystery. Christian Anfinsen received the 1972 Nobel Prize in chemistry for showing that the sequence of amino acids on a protein chain sets the broad structure of a molecule embedded in a cell's watery solution. The amino acids that are water-soluble try to move to the outside of the chain, while those that are water-insoluble attempt to stay away from the water. The folding up, guided simply by thermodynamics, attempts to line up the first kind of molecules on the outside and the second on the inside. It's a beginning, but more is required. That's where the heat shock protein, hsp, comes into play.

The mechanism works a bit like an automobile body shop. If a molecule is bent out of shape by an accident, the tow truck, hsp 70, grabs hold of it and brings it to the shop, hsp 60, where little hsp 10 tools work on it and send it back out to the cell. Hsp 60 is shaped like two rings, one on top of the other—a convenient resting bench for a stressed molecule. Heating increases the likelihood of an accident because molecules move faster as the temperature goes up. The faster the long, carefully shaped molecules move, the more likely they are to be bent out of shape in an accident. I like the automotive analogy, but most scientists call hsp 70s "molecular chaperones," perhaps reflecting the role they play in accompanying molecules to the repair sites.

Drosophila led the way to uncovering a secret, but the production of hsp 70 is universal across species. By the end of the 1970s, similar proteins were found in bacteria, plants, and animals, always as a response to increases in temperature. In *E. coli,* abundant hsp production sets in by 100 degrees Fahrenheit; by 120 degrees, heat shock proteins are the only proteins being made. The *E. coli* are dying but heat shock proteins are still trying to do their work. If this kind of repair work was all heat shock proteins did, it would be interesting, but there's more to the story.

Rising temperature is only one of many environmental stresses that lead to the deforming or improper folding of molecules in the cell. Poisons, heavy metals, and pollutants of one sort or another are often just as bad or worse. As the 1970s came to a close, researchers found that these cellular invaders also generated hsp production in the cell in much the same way that rising temperatures do. The stress response is so universal that heat shock proteins are now often simply called stress proteins.

These proteins are believed to play an important role in human disease. For instance, the immune system operates by recognizing invaders, attacking back, and destroying them, but there may be an intermediary step in which the invader triggers stress proteins that alert the immune defense. It seems likely that these proteins also play an important role in the human fever response. Perhaps the rise in temperature during disease is simply one way of stimulating our body to increase its hsp production. A response as basic as hsp production, shared by creatures ranging from fruit flies to humans, is bound to be part of the explanation for fever.

In the second half of the nineteenth century, humans realized they were related to the great apes. It was a first great step toward recognizing the communality in life's adaptive mechanisms. The mid-twentieth century showed us that all living creatures employ the same procedures for using DNA and RNA to encode genetic information, in what Francis Crick calls the "central dogma of molecular biology." Nevertheless, we at first clung to the belief of separateness by thinking that the information contained in genes was different from species to species. But genetic sequencing in the latter part of the twentieth century has proved how similar we are not only to apes, but to frogs, sea urchins, fish, and even to yeast cells. I think one can say the realization of this communality has been the greatest insight of the last twenty years in biology.

For example, the Hom gene establishes the dorsal structure of the fruit fly, distinguishing anterior from posterior alignments. Astonishingly, geneticists have found the same genes in worms, leeches, lancelets, mice, and humans. Furthermore, the gene from a mouse can be inserted in a mutant fly and the fly functions perfectly. There is also a master gene for vision, almost identical in the fly and the mouse. Of course the lensed eye of a mammal is totally different from the compound eye of an insect, but the gene that starts the process of developing the function of sight, the root of the activity, is essentially the same.

Temperature regulation is harder to pin down than structural orientation or even sight. There probably is no master gene we can associate with controlling hot and cold, and yet it is a basic function common to all species. What is the root of this control, how is it set, and how is it varied? How common is it to all species? These are very basic questions that we are just beginning to address—great problems to study in the coming decades.

In parallel, we continue to learn how the brain, nerves, skin, blood, and sweat glands dispersed through our bodies maintain our uniform temperature and how other species set their own commands. The study of communality and the appreciation of the details of diversity are moving forward side by side, as they must.

While we are similar to even the simplest organisms, there are crucial ways humans are different from all other animals. We read, we write, we make fires, and we even measure how hot they are.

Chapter 2

MEASURE FOR MEASURE

BEFORE 1600, people didn't think much about measuring temperature other than following rough prescriptions for what to wear, where to go, how to cook, and how to forge a tool. Admittedly, Galen tried to give guidelines for medicinal administrations, dividing into four measures the deviations of hot and cold each way from a standard. There was also common knowledge of how much wood was needed to boil water or fire a pot, but such skills were acquired simply through practice. By contrast, the standard measurements for length, time, and weight were very precise since these were needed for the orderly function of government, commerce, and everyday life. You can find the metric equivalent of the Sumerian *zir,* the Akkadian *ubanu,* the Assyrian *imeru,* or the Jewish *gomor,* but you will not discover records of temperature measurements. There were none.

The ancient Greeks were remarkably skillful at deducing lengths, even ones they couldn't measure directly. In the third century B.C.E., Eratosthenes estimated the radius of the Earth to within 5 percent of its true value by observing the angle between the noontime Sun's rays and a vertical stick at two widely separated cities. In the third century B.C.E., Aristarchus assessed the size of the Sun and of the Moon as well as our distances to them.

Time was harder to measure than length, but increasingly

elaborate sundials, water clocks, and sandglasses worked well. The flow, or rather the trickle, of a fixed amount of water from one receptacle to another, corrected for evaporation, gives an accurate and consistent measurement of time. Days and years were set by astronomical readings. Ships left at appointed hours and weddings were conducted at agreed upon dates, but still, no records of temperature. While time is the conventional measure we use for the evolution of the universe, of the Earth, of life, of human existence, of civilization, it isn't always the best measure. Early cosmology, as I'll show later, commonly uses temperature rather than time to mark the unfolding of events in the cooling universe.

If I were to employ temperature as my record for a narrative of civilization, I would cite the ever-hotter fires humans made as they moved from hunter-gatherers to villagers to toolmakers—from the Stone Age's first fires to charcoal and then to the bellows that produced bronze and iron. Going further, I would reach the steam engine, the nineteenth century's great Bessemer furnaces that made steel, and finally the Nuclear Age. My history would read 0 degrees, 500, 1000, 2000, 2500, and finally millions of degrees Fahrenheit. For the past 200 years, I could use as a marker the ever lower temperatures achieved in laboratories as, one by one, all known species of gas were liquefied. As the millennium came to an end, the low-temperature scale reached billionths of a degree above absolute zero.

The First Sparks

We don't know which, if any, of the ancestors of *Homo sapiens* first mastered the trick of making a fire or when the first flames went up. We do know that we are all descendants of southern apes of the *Australopithecus* genus. Our direct ancestors, emerging in central-southern Africa, apparently split away from chimpanzees some five million years ago, moving on their own line of evolution from *A. anamensis* to *A. afarensis* to *A. africanus*. This last ancestor was about four feet tall, weighed some sixty pounds, had a brain size one-third of ours, and probably looked more like the present-day chimpanzee than a human. Pelvic modifications already allowed it to walk erect rather than dragging its knuckles. Then, about 2.5 million years ago, another split occurred. Some of the early ape species became extinct, but one survived, evolved, and thrived, mutating from *Homo habilis* to *Homo erectus* and finally to *Homo sapiens*.

Early caves have chemical evidence of wood fires from at least 200,000 years ago, the period of the transition from *Homo erectus* to *Homo sapiens*. These caves are marked by the accumulation of ashes containing minerals characteristically absorbed by trees. This suggests the wood was burned, but that isn't enough to prove the existence of human fires or even of human presence. We look for abundant stone tools, primitive hearths, and large mammal bones, the remains of animals that were hunted, cooked, and eaten, assuming, of course, humans were already carnivorous. Such a site was found in Hungary near Vertesszollos, with the burned bones even arranged in a radial pattern such as one might find in a campfire. Those were clearly made by human fires and date back more than 200,000 years. The chronology is unclear further back.

For the last fifty years archaeologists thought the oldest historic site with indications of manmade fire was Zhoukoudian, about thirty miles southwest of Beijing. Supposedly a *Homo erectus* cooked a deer there 500,000 years ago, but a recent reanalysis of the excavation suggests the burned bones are too scattered to have come from a single campsite. There is also little evidence of fireplaces or hearths. The remains at the site appear to be organic material, but not the kind of ash generated by burning wood. The upshot is we still have no clear evidence that any predecessors of *Homo sapiens* made fires.

In looking for bones in old campsite hearths, archaeologists generally assume that early fires were used to cook meat. The traditional view is that the emergence of *Homo sapiens* paralleled a switch from a diet of fruits and berries to one rich in meat from scavenged and perhaps hunted animals. Marrow protein could have triggered growth in size and body development. But Harvard anthropologist Richard Wrangham and his colleagues claim the change in diet triggered by the use of fire did not involve shifting from a vegetarian to a meat diet. They hypothesize it was acquisition of the ability to dig up and cook tubers like cassava and yams. Going a step further, they suggest this new diet led to wholesale changes in family and tribal structure, revolving around the gathering, storing, and processing of tubers. Raw tubers are stringy and indigestible, but cooking turns them into easily consumable calories. In modern-day Africa, tubers are still an abundant part of the diet. Wrangham says their starchy roots were being cooked on the plains of Africa 1.8 million years ago. The claim, one of many competing theories of human social evolution, is certainly being questioned. Fires where meat was roasted leave better remains—burnt bones—than fires where only roots were cooked. Nevertheless, roots should have left some clear chemical traces.

Beyond diet, fire changed the way people lived. A blazing bonfire scares animals away. A fire at the entrance to a cave keeps out even the most determined predators. Fire also provides light in that cave, enabling habitation of its deeper reaches. Fire meant that humans could move from the temperate zones in which they had evolved toward harsher climates, particularly if they had tools to help them hunt and make clothing.

Sophisticated weapons and tools appeared in both southeastern Europe and the Near East around 50,000 years ago, a time that Jared Diamond calls "The Great Leap Forward." Hunting large animals from afar using spears and harpoons became possible and sewing improved the quality of clothing. Art was born, as the great wall paintings in the Lascaux caves testify. In spite of these advances, humans still remained in the Stone Age. The tools and weapons became more sophisticated but were still shaped by pounding with rocks or were flaked from other rocks. Temperature was still not a factor in tool making.

Shortly after this transition, humans began taking clays, plentiful in the Earth's crust, and treating them by drying and baking in simple fires. Aside from hardening, clay will also acquire color in the process of firing because of the iron oxide usually present. This brightens its reddish hue when fired in an oxygen-rich atmosphere and converts it to blackish magnetite when fired in a smoky, oxygen-poor flame. The oldest known clay figurines, dating back 27,000 years, were found in what is the present-day Czech Republic; the oldest fired clay pottery, found in Japan, is 14,000 years old.

Tools for harvesting wild crops appear in sites from around 11,000 B.C.E. Some humans began to make the dramatic shift from hunter-gatherers to farmers around 10,000 years ago. This happened along the Mediterranean, in Southeast Asia, and the Fertile Crescent, the region in the Middle East around the Tigris and Euphrates rivers. By 2500 years later, clear indications exist of wheat and barley having been raised in the Fertile Crescent. The need for better tools for farming, better containers for storage, and better weapons for war all pushed these early civilizations to improve fire making. This is particularly evident in the working of metals.

The first worked metals were soft ones like gold or copper that were malleable and could be hammered as is—without being heated up—into sheets and then shaped by the use of stone tools. By around 3000 B.C.E., it was discovered that if you heated and mixed copper with smaller amounts of other metals, chiefly tin, it led to an alloy that was stronger and more durable. We call it bronze. The Greeks called it *chalkos,* the Romans, *aes.*

Bronze was historically a smelted mixture of copper and tin in roughly an eight-to-one ratio, but more copper gave a softer metal, better adapted to shaping utensils and armor while less copper yielded a harder one, more suitable for swords or bells. As time went on and societies became more sophisticated, other metals were added to the mix in large or small amounts as the needs and availabilities changed. Bronzes have been made with zinc, lead, silver, antimony, arsenic, aluminum, and phosphorus, all elements that alter the properties of the alloy, either for functional or decorative purposes. To give one example, Hephaestus in the *Iliad* is described as making Achilles' shield by throwing into his cauldron copper, tin, silver, and gold. This era of bronze tools, used for domestic purposes and warring, is one that most civilizations have passed through. It is so important, so pronounced, and so characteristic that archaeologists have simply named it the Bronze Age.

The Stone Age, Bronze Age, then the Iron Age. Iron implements are much harder than bronze ones and, with the production of iron, a whole new range of possibilities arose. Civilization changed again. The Iron Age is marked by fire temperatures that reached between 2000 and 2500 degrees Fahrenheit. It began in the Fertile Crescent almost 2,000 years after the start of the Bronze Age, while in other parts of the world the Bronze Age had a later onset and a shorter duration.

Iron smelting, which comprises two steps, already requires a good deal of sophistication beyond simply making fires. The first step in purifying iron, normally found in the earth as iron oxide, is the stripping away of the oxygen. Known as reduction, this is performed at 1500 degrees by actually adding oxygen into the furnace. The free oxygen combines with carbon from burning charcoal to form carbon monoxide. This removes oxygen from the iron oxide, making carbon dioxide in the process. The carbon dioxide is carefully vented out, leaving iron behind. This primitive iron still contains many impurities, however. In the second step, after further heating, the molten iron forms a slag that floats off, leaving pure iron. In this final stage, temperatures must reach well above the melting temperature of the impurities but well below that of iron, 2800 degrees. I mark this stage of civilization by about 2500 degrees Fahrenheit.

As I hinted earlier, the transition from stone to bronze to iron with hotter fires is paralleled by changes in the firing of clays. Iron oxide is plentiful in the earth, but so are the microscopic aluminosilicate particles that make up clays. They are layered so as to slide easily over one another, allowing shaping of clay objects. When fired, clay melts just enough to solidify into a hard mass while still retaining its porous quality. To counteract the porousness, fired clay is often sealed by a glaze composed of glasslike particles that melt at lower temperatures and can be applied without altering the shape of the underlying vessel. Mixed with a little water, the glaze is applied to the clay. In the end, the impurities left in the glaze give it the distinctive properties we admire. As has happened so often in human history, a commodity becomes an art form in the hands of a skilled artisan. The mixture of the glaze, the time of firing, the impurities in the glaze, and the atmosphere in the kiln all add to the quality of the clay's covering.

Godin Tepe in western Iran is the oldest known site where glazed pots have been found. Analysis shows those glazes had been prefired at 1900 degrees Fahrenheit, then ground up, applied to a pot, and fired at almost 1500 degrees. This technique seems to have disappeared until 1500 B.C.E. when it reappeared on a large scale in China. The great pottery art of the Chinese is due to many factors, but among them are the abundance in China of excellent clay deposits and limestone that make good glazes. By 1500 B.C.E., pottery was routinely being fired at 2000 degrees.

In Greece, by 500 B.C.E., glazes were rendered liquid and painted onto great clay receptacles that were fired in a smoky atmosphere, turning the whole vase black. As an atmosphere richer in oxygen was added to the kiln, the unglazed clay turned back to red, leaving the great Attic vases that we admire so much.

Fires made by humans reached 2500 degrees about 2000 years ago. For the next 1600 years, however, not very much progress was made in fire making, in medicine, in astronomy, and in measurements as a whole, at least in the Western world. But then things changed, in part because of remarkable advances in how temperature came to be measured.

The Thermometer's Four Inventors

The early seventeenth century was a prolific period in the invention of measuring devices: during this time the telescope, the microscope, the thermometer and, a little later, the pendulum clock were created. However, measurement is only scientifically significant if the observation contributes to proving or disproving something. Thus, fame goes rightfully to the one who understands the significance of the measurement, not the tool's inventor. Nevertheless, such scientific observations and discoveries would be hard to come by without the existence of these instruments.

The history of the thermometer resembles that of the microscope more than that of the telescope in the sense that its invention did not instantly change our view of the world and no great discoveries were quickly made with it. I'll compare the three.

The first telescope was probably made around 1600 by one of several Dutch spectacle makers. Hans Lippershey, James Metius, and Zacharias Jansen all have claims, but in a deeper sense the telescope was invented when Galileo pointed it at the sky on January 7, 1610. While focusing on Jupiter, he saw what he thought were four little "stars" near it. Nothing surprising about that, since Galileo was discovering hundreds of stars; he was, after all, the first to observe that the Milky Way was simply a collection of many distant stars. On January 8, the four stars had moved relative to Jupiter. January 9 was cloudy. On the following day, when it was clear again, Galileo observed a new change in the position of the four stars. Others would have said this was an insignificant event, but Galileo, with his usual sureness, realized that these were not stars. They were moons orbiting Jupiter. Each completed the circuit around the planet in a few days. He describes the phenomenon, still using the word "stars" instead of "moons."

> But what surpasses all wonders by far, and what particularly moves us to seek the attention of all astronomers and philosophers, is the discovery of four wandering stars not observed by any man before us. Like Venus and Mercury,

> which have their own periods about the sun, these have theirs about a certain star that is conspicuous among those already known, which they sometimes precede and sometimes follow, without ever departing from it beyond certain limits.

In 1600, you could still reasonably deny Copernicus and believe the Earth was the center of the universe, with the Sun and all the planets revolving around it. There simply was not enough clear-cut evidence to the contrary. Those four moons, however, revolved around Jupiter, not around the Earth. One of the objections to the Copernican view was that the Moon could not follow the Earth if the Earth rotated around the Sun. This was, however, another example of moons rotating around a planet just as our own Moon rotated around the Earth. If Galileo's interpretation was right, the Earth was not the center of the universe.

I like to think of Galileo's observation of the moons of Jupiter as one of the beginnings of modern science. There was a model, the Copernican one, which Galileo had believed in because he thought it provided the best explanation of existing data. He then looked for more evidence to prove or disprove the model. Finding that evidence, he realized what his discovery meant for the model. The first optical astronomer was clearly a genius, but Galileo's proposal still awaited confirmation or disproof, and he had the right tool for the search.

The microscope was also invented by Dutch lens makers at the beginning of the seventeenth century, but there was no testable model for the nature of microscopic life, so it was little more than a curiosity, a window on a world invisible to the naked eye. Robert Hooke was the first great popularizer of the new instrument. In 1665 he published *Micrographia,* a book that became a best-seller in Europe. In it, Hooke presented fifty-seven illustrations drawn by him of the marvels he had seen with the microscope. These included the eye of a fly and the anatomy of a flea. He also studied a simple cork under the new microscope and stated that it was made up of substructures that he called cells. Hooke sounded, as Daniel Boorstin puts it, the keynote of the new age with his statement, ". . . the science of Nature has been already too long made only a work of the brain and the fancy. It is now high time that it should return to the plainness and soundness of observations on material and obvious things." Nevertheless, the microscope remained only a curiosity, without an application or a purpose.

Hooke was a remarkable man whose fame would be considerably greater were it not for the fact that he is completely overshadowed by Sir Isaac Newton. On reading Newton's *Principia,* Hooke claimed the idea of the inverse square law for gravitation had been stolen from him. Hooke did have the idea independently of Newton, and even realized that an object falling toward Earth was the same kind of motion as the Earth's falling toward the Sun, but he lacked Newton's astonishing mathematical genius, which allowed him to derive Kepler's laws of planetary motion from those assumptions. Newton, no mean quarreler himself, as we know from his disputes with Leibniz about the discovery of calculus, promptly removed all mention of Hooke from the *Principia* and refused to have anything more to do with the Royal Society, Hooke's employer, agreeing to become president only after Hooke's death in 1703. When Newton did become president, Hooke's portrait hanging in the Royal Society mysteriously disappeared.

Now for the thermometer. Heat had been studied before 1600: Philo of Byzantium considered the expansion and contraction of gases in the second century B.C.E., and about 200 years later, Hero of Alexandria wrote a book called *Pneumatics*. When translated into Latin in the Renaissance, the book was rediscovered and went through repeated Italian editions. Galileo read it in 1594. One of Hero's experiments involved a device called a thermoscope, intended to show the change in a gas as it was heated or cooled. For instance, a sealed container of air topped by water was heated, causing the gas to expand and the water to rise. This kind of phenomenon is connected in some ways with temperature but not necessarily associated with any measurement.

Ironically, Galileo, who rightly gets so much credit for the telescope even though he did not invent it, also is usually credited with inventing the thermometer, although he hardly used it at all and certainly not for anything notable. Galileo's claim is shared by at least three other individuals, probably because, as with the telescope, several individuals had more or less the same idea at the same time.

Two of the potential inventors of the thermometer were from north of the Alps and two were from Italy. One of the two northerners was a Welshman named Robert Fludd, who settled in Oxford, took a medical degree there, and possibly built a thermometer of sorts after reading, in Oxford's Bodleian Library, a thirteenth-century manuscript describing Philo's thermoscope. The other was a Dutch inventor named Cornelius Drebbel; he moved to England and was employed by King James I as a kind of court inventor. Performing such feats as the apparent cooling of Westminster Abbey in midsummer, he built several devices, including perhaps a thermometer.

Galileo was, of course, one of the Italian inventors, and the other was a friend of his named Santorio. Three years older than Galileo, Santorio Santorio was born in 1561 to a wealthy Venetian family which, according to a custom at the time, gave him the family name as first name. He studied philosophy and medicine in nearby Padua and eventually returned to Venice to practice medicine. Venice at the time had already embarked on a gentle decline from its peak as the dominant naval power in the Mediterranean and the gateway to commerce with the Orient, but it was still a cultural and artistic center. Nearby Padua housed the Venetian Republic's great university. The city's relative distance from and animosity toward the Papal States gave Venice and its surroundings a degree of intellectual freedom less easily obtainable in other parts of Italy, as Galileo sadly discovered when he left Padua for Florence.

Santorio set out to quantify humoral medicine. He was going to take Galen's 1500-year-old notions of hot and cold, wet and dry, bile and phlegm and bring them up to date. Left unchallenged, they had never been exposed to the kind of rigorous study through which Kepler had deduced the laws of planetary motion. Santorio was going to use the new scientific instruments to make a science of humoral medicine. Experimenting on his own body, Santorio carefully weighed himself before and after eating, drinking, sleeping, and exercising. He recorded his intake and his excretions, attributing the difference between the two to "insensible perspiration." He also kept a diary of his temperature variations.

Santorio's *Ars de Medicina Statica,* first published in 1612 and translated into many languages, was widely circulated throughout Europe. It came to be thought of, for over a hundred years, as one of the two pillars of modern medicine, Harvey's book on the circulation of blood being the other. We now realize that Harvey's insight into the flow of blood in the body was brilliant and Santorio's insights on humors were not. While temperature and pulse rate are still regularly recorded on hospital charts, as I said in the last chapter, they do not denote imbalance of humors. Nevertheless, though Santorio's idea was flawed, his methods were ingenious. For instance, using Galileo's discovery that a pendulum of fixed length always swung with the same period, he adjusted a pendulum's length until the period coincided with a patient's pulse rate and then measured deviations.

In the end, Santorio's attempt to "scientifically prove" humoral medicine failed, but along the way he laid the foundation for the study of metabolic activities. He took a thermoscope much like the one Hero of Alexandria had described over 1500 years earlier and put a scale on it in order to measure deviations from normal body temperature. A thermoscope with a scale on it is basically a thermometer; it measures in a quantifiable way the divergences of temperature from an accepted point. Santorio deserves some credit for building this device and, in any case, is the first person to systematically measure body temperature.

Meanwhile, Galileo's biographer, Viviani, claims the thermometer was invented in Galileo's early days in Padua, between 1592 and 1597. As Viviani puts it, "In these same years he discovered thermometers, that is to say those instruments of glass, with water and air, for discerning the changes of heat and cold." My guess is that, after reading Hero's *Pneumatics* in 1594, Galileo tried to build a thermoscope of his own and doubtless improved on the original suggestion, just as he had done with the telescope. He also probably put a measuring scale on it. A 1613 letter to Galileo from his friend Sagredo, who was experimenting with thermometers, also mentions that Galileo invented them.

The modern thermometer, still used despite the recent dominance of digital instruments, has liquid sealed in a glass tube with a scale etched on its side. Mercury, employed as far

Fresco depicting a meeting of the Accademia del Cimento in the seventeenth century

back as 1700, is the common fluid in the tube, though early instruments contained alcohol, various spirits, and even water, all with or without coloring. The instrument was probably first developed in Florence around 1640. Credit for the invention is usually given to Ferdinand II, the Grand Duke of Tuscany, to whom Galileo's *Dialogue concerning the two Chief Systems of the World, Ptolemaic and Copernican* was dedicated. When Galileo died in Florence in 1642, Ferdinand called him "the greatest light of our time" and sought to have a marble mausoleum built in his honor in Florence's Santa Croce, traditionally a burial place for the great. Pope Urban VIII vetoed the construction, saying it would be an offense to papal authority.

Despite Galileo's problems with the Inquisition, a lively scientific community continued to flourish in Florence, partly because of the benevolent patronage of the Medici family and partly because of the presence of a number of disciples/students of Galileo's. These include Cavalieri, Viviani, and Torricelli, the inventor of the barometer. In 1657, a group of nine Florentines joined together to form a learned society, the Accademia del Cimento. The members, following Galileo's teachings, built their own apparatus and conducted their own experiments to advance the pursuit of knowledge. After ten years, they published an account of their work entitled *Saggi di naturali esperienze fatte nell' Accademia del Cimento,* with *saggio* meaning a kind of report. The first of these *saggi,* entitled "Explanation of some instruments for finding out the changes of the air resulting from heat and cold," gives the details of making a thermometer, including a description of how the glassblower makes the bulb.

Some of these early thermometers, on view in Florence's Museo di Storia della Scienza, are extraordinarily beautiful. Contained in glass blown into a variety of shapes—tubes, spi-

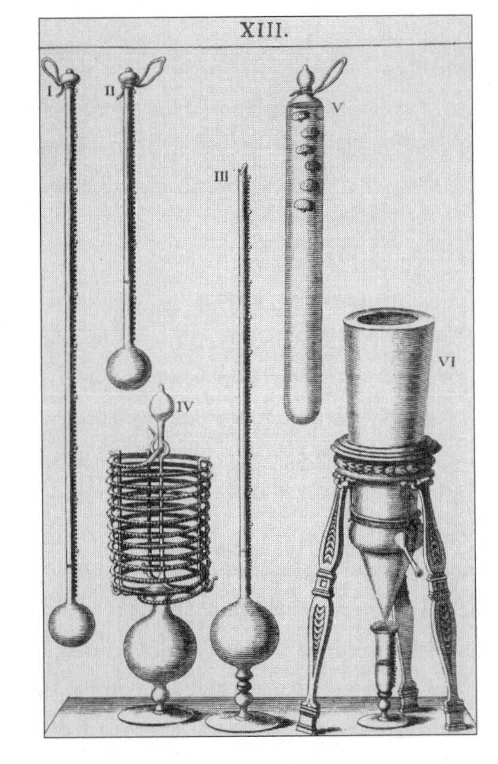

Early thermometers and a hygrometer constructed by the Accademia del Cimento

rals, and even fantastic animal shapes—the thermometers all operate on the basic principle that the enclosed liquid will expand when heated. That expansion is noted by elaborate markers or by small free-to-move spheres of varying density contained in the thermometer, sensitive to the liquid's change in density as temperature varies.

Unfortunately, the Accademia was disbanded in 1667, only ten years after its founding. The termination of the society appears to have been a condition for the appointment of its patron Leopoldo de' Medici as cardinal, and it marked the decline of free scientific research in Italy, which had already been dealt a serious blow by the Catholic Church's trial of Galileo for his supposed defense of the Copernican view of a heliocentric universe. There were certainly other causes for the decline, among them Italy's division into many smaller states, but scientific history might have looked somewhat different if the Accademia had thrived.

The subsequent history of the thermometer is interesting, but has little to add to the underlying science. Hooke built some thermometers and Newton did as well. In 1701 Newton suggested using linseed oil as a fluid and introducing a 0–12 scale, the limits being what is found in "compressed snow as it is melting" and "the maximum heat that the thermometer can attain by contact with the human body." The usually secretive Newton didn't even reveal he was the author of this work; others guessed it and he finally confirmed their surmise.

A few years later a Dane named Ole Romer (principally famous as the first to measure the speed of light) suggested that the melting point of ice and the boiling point of water would make a better scale. He set these at 7.5 and 60. Romer was visited by an instrument maker named Daniel Gabriel Fahrenheit, who took up his idea, eventually changing the scales again. They are now set at 32 and 212. The reasons for these two seemingly arbitrary numbers are probably that Fahrenheit was setting his zero by a mixture of water, ice, and salt and his hundred near body temperature. The Fahrenheit scale caught on in England and Holland, but most countries, though adopting the freezing and boiling points of water as markers, have adopted the "centigrade" scale, now known as the Celsius scale, after its pioneering advocate, Anders Celsius. This scale is more in line with the general division of units into decimals that came about with the adoption of the metric system after the French Revolution.

Measurement is only interesting when you know what you are measuring—when there is a purpose. We know what zero length or zero time is, but what does zero temperature mean? Even though it is easy to tell which of two objects is hotter, at least if they are in the range of our neural sensors, temperature seems to lack the absolute scale that is so intuitively clear for both length and time. As people began to measure temperature in a "scientific" way, it also became natural to ask what was one measuring. That required a better understanding of heat.

The Count from Massachusetts

Robert Boyle, the fourteenth child of the first Earl of Cork, performed the first set of experiments that begin to connect heat, temperature, and energy. Boyle settled in Oxford in the 1650s, a time when the notion of experimental scientific research was just beginning to take hold in Britain. Independently wealthy, Boyle began doing experiments and discussing their results with a small group of like-minded individuals. In 1663 they joined together to form the "Royal Society of London for improving natural knowledge." He also saw to it that this Society, now commonly known as the Royal Society, hired a young man to be curator of experiments, a position that carried with it the duties of assisting members in carrying out experiments and in preparing demonstrations for the Society's regular meetings. Boyle's choice for the position was the twenty-seven-year-old Robert Hooke, who had assisted Boyle in building a variety of vacuum pumps. Unlike the Accademia del Cimento, the Royal Society thrived.

Boyle's experiments showed that, at constant temperature, the pressure on a gas and the volume of that same gas were inversely proportional to each other. As one went up, the other went down, but their product remained the same. Since a gas also expanded as its temperature rose, this suggested that the temperature of a gas was related to pressure and volume. Approximately a hundred years later, a Frenchman named Jacques Charles took the next step by noting that at constant pressure, the volume of gas in a balloon was directly proportional to temperature. In other words, volume went up or down in exactly the same way temperature did.

In 1802 John Dalton and Joseph Louis Gay-Lussac came up with a very interesting conclusion (incidentally, Gay-Lussac rediscovered by accident Jacques Charles's unpublished and little-known work, and then went out of his way to ensure that Charles received credit for it). It seemed that, once pressure was kept fixed, near zero degrees Celsius all gases increased in volume by $1/273$ the original value for every degree Celsius increase in temperature, and correspondingly decreased by $1/273$ for every degree decrease in temperature. At 10 degrees, the volume would become $283/273$ of its original value, and at -10 it would be $263/273$ of that same original value, having jumped, as they observed, by $1/273$ for every 1-degree shift.

Gay-Lussac extended this relation by showing that, when volume was kept fixed, gas would increase or decrease the pressure exerted on the outside of the gas container by the same 1⁄273 factor when temperature was shifted by a degree Celsius. This was becoming more puzzling. A gas's pressure, volume, and temperature seemed to be related in a way that did not depend on the kind of gas being studied, hinting at some deep connection shared by all gases. Gay-Lussac checked that this wasn't a peculiarity of the gases he used or of where he did his experiments. He was both careful and adventurous in his investigation: in September 1804, he ascended up to 23,000 feet in a hot-air balloon. He took air samples and recorded temperatures as he went up and down. His formulas still held.

If Charles's, Dalton's, and Gay-Lussac's results were correct, there was a startling conclusion. If the volume of gas at fixed pressure decreased by 1⁄273 for every 1-degree drop, it would reach zero volume at –273 degrees Celsius. The same was true for pressure at fixed volume. Since pressure and volume are both positive quantities, zero is the smallest value they can take. Therefore, the drops in volume and pressure had to stop at –273 degrees Celsius. That had to be the end of the scale, the lowest possible temperature one could reach. It was nothing less than absolute zero.

This is the first point I know of at which the idea of an absolute zero temperature enters science. Furthermore, remarkably enough, the notion is entirely correct, and –273 degrees is even the right value (technically, it is –273.16, but let's not quibble). However, none of the scientists involved saw any special significance in the number –273. The extrapolation in temperature to that point was fascinating, and certainly indicated common features of gases, but they simply assumed that all gases would liquefy well before then, so absolute zero wasn't especially interesting to study or ponder.

An impasse of sorts had been reached. Any further

progress on the significance of temperature had to await a better understanding of the nature of heat: was it a separate essence of sorts or wasn't it? In a debate that continued well into the nineteenth century, many considered heat to be a substance present in combustibles and then released upon burning. It was known as phlogiston, which is Greek for "combustible." By 1800, after Lavoisier's experiments proved heat to be weightless, it was known as caloric. A second school, already advocated by Newton and closer to the present view, thought heat was due to motion of an object's constituents, though the nature of these constituents continued to be a mystery.

Some results apparently contradicted caloric theory. In an 1807 gas experiment, Gay-Lussac took a large container with a removable divider down the middle and filled half with gas and made the other half a vacuum. When the divider was suddenly removed, the gas quickly filled the whole container. Since, according to caloric theory, temperature was a measure of the concentration of caloric fluid, removal of the divider should have led to a drop in temperature because the fluid was spread out over a greater volume without any loss of caloric fluid. (The same amount of fluid in a larger container means lower concentration.) Gay-Lussac found, however, that the temperature didn't change. Calorists tried to explain this by saying the fluid had been damaged or altered in the expansion.

Evidence linking heat to mechanical energy was accumulating. Expenditure of the latter seemed to lead to the former. Credit for establishing a definite connection between the two, the so-called mechanical equivalent of heat, is usually given to a Count Rumford, who wrote "An Inquiry Concerning the Source of the Heat Which is Excited by Friction," which appeared in the 1798 Transactions of the Royal Society.

Count Rumford, born in Woburn, Massachusetts, in 1753 and christened Benjamin Thompson, was not what one usually thinks of as a count. At nineteen he married a thirty-three-year-old wealthy New Hampshire woman, the widow of a colonel in the New Hampshire militia. The governor of New Hampshire approved the marriage and quickly acted to make young Thompson an appropriate spouse by appointing him a major in the same militia in which the widow's previous spouse had served. Favoring the British cause as the American Revolution began, Thompson abandoned his wife and daughter and moved to London, eventually serving as undersecretary of state for the colonies. While so engaged, he simultaneously pursued a scientific career, at first directed to the needs of the British army. Elected to the Royal Society for his research on explosives, Thompson is also known for his numerous inventions, including new kinds of stoves, the double boiler, and the drip coffeemaker.

After being knighted by George III, the now thirty-one-year-old Thompson went to the Continent to seek his fortune. An attractive offer by the Elector of Bavaria led him to settle in Munich. He introduced Bavarians to Watt's steam engine, planned Munich's English Gardens park, and finally was made Count Rumford of the Holy Roman Empire in gratitude for his service reorganizing the army and civil service (he took the name Rumford from the original name for Concord in New Hampshire, where his career started). Rumford eventually moved back to Britain, reentering the country as Bavaria's Minister Plenipotentiary to the English Court. George III refused to recognize one of his own subjects as a foreign ambassador. In a huff, Rumford left Britain and settled in France, which was then at war with Britain. At first he was feted, even marrying the widow of the great Lavoisier, who had been executed in the French Revolution. Eventually,

Rumford's quarrelsome nature took over, and by his death in 1814, he seems to have been at odds with everyone.

This colorful figure is now mainly remembered for a set of experiments he performed during his stay in Bavaria. While perfecting the local armaments, the newly minted Count Rumford began to think about all the heat generated during the reaming out of the metal core when the bore of a cannon is formed. He realized the metal shavings had the same thermal properties as the original metal: it took identical amounts of heat to raise by 1 degree either a gram of shavings or a gram of solid metal. Clearly the metal hadn't changed. So where did the heat come from? Rumford thought the source had to be the work done in drilling the hole. In his view, heat was not an indestructible caloric fluid, as Lavoisier had argued, but rather something that could come and go. Mechanical energy could produce heat and heat could lead to mechanical energy.

Rumford correctly thought there was no separate caloric fluid and that heat was associated with motion or internal vibrations. The analogy he drew was to a bell: heat was like sound, with cold being similar to low notes and hot, to high ones. Temperature was therefore just the frequency of the bell. A hot object would emit "calorific rays" when struck, while a cold one would emit "frigorific rays"—a point of view that goes all the way back to Plutarch's *De Primo Frigido*. Heat was not used up any more than a bell was by ringing. Cold was an entity of its own, certainly not simply the absence of heat.

Using sound as an analogy, Rumford described an experiment to his colleague Raoul-Pierre Pictet in Geneva by saying that "the slow vibrations of ice in the bottle cause the thermometer to sing a low note." According to Rumford, thermal equilibrium between two objects at different temperatures was reached by the reciprocal absorption of frigorific rays by the hot object and calorific by the cold. He even thought people in northern climates were white while those in the tropics were black because white reflected frigorific rays while black reflected calorific ones, a principle he carried out in his own life by always dressing in white in winter.

Even after Rumford's experiments on the mechanical equivalent of heat, there was a long road to travel in understanding heat. Science is usually taught as a kind of triumphal march, from one brilliant result to another. Historians and sociologists of science often criticize teachers for this kind of instruction, but those of us who teach science usually feel there is so much to learn and so little time. We take the simplest route. I often have wished I could do otherwise, because it's instructive to see individuals groping toward an answer, to observe how easy it is to set off on the wrong path. Harvey is remembered for his notion of the circulation of blood, but Santorio is forgotten. Rumford is remembered for his evaluation of the mechanical equivalent of heat and the correct association of heat with the motion of the constituents, but a slightly deeper probe shows the shakiness of his work's underpinnings.

Steam Power

While the debate about the nature of heat continued, the practical association between mechanical and thermal energies was ushering in the Industrial Revolution through the introduction of the steam engine. First built by Thomas Newcomen in 1712 and later perfected by a series of Englishmen, these machines changed both the nature of manufacturing and the transportation of manufactured goods.

Manchester, England, was a town with fewer than 10,000 inhabitants at the beginning of the eighteenth century and was still a small town in 1769, the year James Watt patented the first efficient steam engine. But it quickly became one of the world's most important cities—the center of an industry in which cotton from America arrived through Liverpool, was woven in Manchester, and then shipped around the world on steamships. The city's population quadrupled in the last forty years of the eighteenth century and then quadrupled again in the first forty of the nineteenth century.

Part of the city's rise was simply a historical accident—a fortunate location plus good planning. Part was due, however, to the religious dominance in the city of Unitarians, barred from Cambridge or Oxford. Manchester Academy opened its doors in 1792 to youths who were to be taught a curriculum in which science figured prominently, a revolt of sorts against the education of the day, which was based on the classics. This shift in education, turning the citizens in more practical directions, happened to coincide with the development of the Industrial Revolution. Manchester honored science and technology, in part because of the intrinsic prestige, but also because the city saw its own growth and prosperity directly related to technological innovations.

The city's first great scientist was John Dalton, the originator of the atomic theory and one of the discoverers of the regularity in the expansion of gases. After 1817, he was the president of the Manchester Literary and Philosophical Society, which was founded in 1781. While the residents of Manchester may not have understood the details of atomic theory, they did know what the steam engine and the mills had done for them and that somehow Dalton was linked to this progress. When he died in 1844, there was a funeral cortege with a hundred carriages stretching three-quarters of a mile. Forty thousand mourners passed by Dalton's coffin lying in state in the town hall.

In an essay entitled "Manchester and Athens," the physicist Freeman Dyson discusses the style of research created in Manchester. He takes his title from a novel by Benjamin Disraeli, later prime minister of Great Britain, in which the hero says, "Manchester is as great a human exploit as Athens." Most of that greatness is due, of course, to the city's grasping the notion that, just as electricity did a century later and computers another century after that, the steam engine changed the nature of commerce.

Despite England's primacy in developing steam engines, the Frenchman Sadi Carnot was the first to understand and formulate their surprisingly simple underlying principles. In doing so he also arrived at what eventually came to be known as the Second Law of Thermodynamics. Carnot, an engineer by training, was deeply concerned with France's decline after the fall of Napoleon in 1815. He came from a prominent French family with strong political and scientific interests; his father had been Napoleon's minister of war, his brother was a prominent politician, and his nephew was president of France from 1887 to 1894.

Carnot realized that England's rise and France's dip in world standing were not just due to the Battle of Waterloo. They were also due to the lack in France of cities like Manchester. As he said:

> Iron and heat are, as we know, the supporters, the bases of the mechanic arts. It is doubtful if there be in England a single industrial establishment of which the existence does not depend on the use of these agents, and which does not freely employ them. To take away from England her steam

> engines would be to take away at the same time her coal and iron. It would be to dry up all her sources of wealth, to ruin all on which her prosperity depends, in short, to annihilate that colossal power. The destruction of her navy, which she considers her strongest defense, would perhaps be less fatal.

In 1824, twenty-eight-year-old Carnot began studying steam engines. He realized they could be viewed schematically as engines that operate in a cyclical fashion. Water boils, changes to steam, and then enters a cylinder that pushes a piston. The piston completes its stroke and returns to its original position. The steam cools and is led to a condenser, and from there the water goes back to the boiler where the cycle begins again.

Carnot argued that the steam engine is analogous to a waterwheel in which falling water drives a paddle. The greater the distance the water drops, the higher the rate at which work is done, or, more simply, the faster the wheel turns. In a brilliant insight, Carnot concluded that the rate at which work is done by a steam engine depends only on the temperature difference between the source and sink of heat, the boiler and the condenser. The temperature difference is the analogue of the height difference from which water falls in a waterwheel.

Carnot did not fully grasp the fundamentals of how a steam engine works, nor is his analogy complete. For instance, Carnot thought heat was indestructible, a conserved caloric fluid: heat taken in equals heat released. This of course fit in with his mental picture of a waterwheel. One might think that Rumford's experiment of changing mechanical energy to heat proved there is no such thing as caloric fluid, but advocates of the fluid point of view took Rumford's work as a confirmation of their ideas. They interpreted his experiment as meaning that friction produces heat by squeezing caloric fluid out of an object.

Shortly after Carnot's tragically early death at age thirty-six (he died of cholera), it became increasingly clear that there is no such thing as a caloric fluid: heat is simply a form of energy, one of many, and the sum of all forms of energy in an isolated system is conserved. This has come to be known as the First Law of Thermodynamics. In the case of a steam engine, the heat taken in at the boiler is not equal to the heat removed at the condenser. The work done by the ideal engine is the difference between the two.

The first "modern" experiment proving the First Law of Thermodynamics was performed by a modest young student of Dalton's named James Joule. Educated in Manchester's practically minded tradition, where experiment rather than preconceived notions ruled, Joule started his research on the topic by showing that an electrical current run through a wire produced heat; this meant electricity and heat were related. Perhaps the electric current was squeezing caloric fluid out of the wire, but Joule didn't think so.

Joule next turned to an improved version of Rumford's experiment. He built a miniature blender whose paddles, immersed in a beaker, were connected to a central rod. A slowly falling weight was connected by a pulley to the rod. In essence, the falling weight turned the blades that stirred the water, without changing the water in any way. Using a very accurate thermometer, Joule was able to show how much heat was created in the water by the change in potential energy of the dropping weight. Repeating the experiment with the blades immersed in different fluids gave the same answer for the amount of heat generated. This proved, at least to Joule,

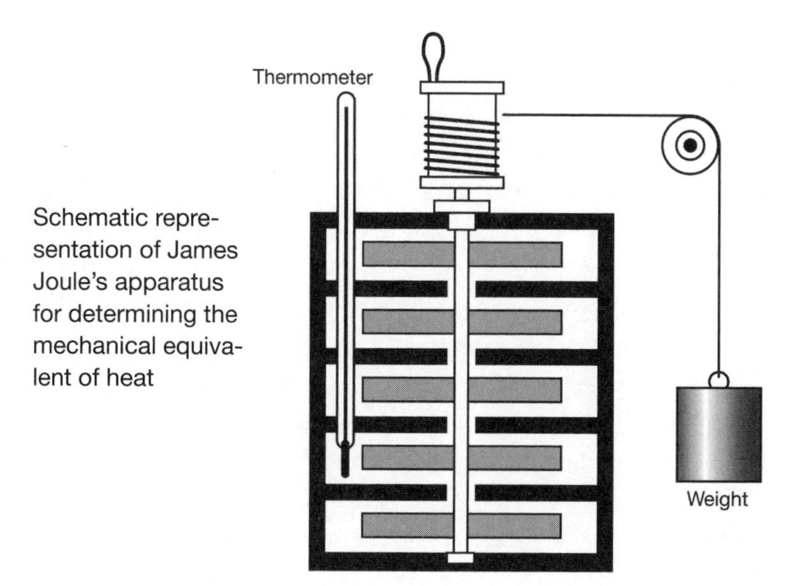

Schematic representation of James Joule's apparatus for determining the mechanical equivalent of heat

that energy is conserved. Thus, there could be no such thing as a caloric fluid.

In 1847 a young Scot named William Thomson heard Joule lecture about his measurements. Thomson was a prodigy, a twenty-three-year-old professor in Glasgow, widely read and very up to date. He was puzzled by Joule's claim because it seemed to contradict Carnot's ideas. These were still unknown in Britain, but Thomson, who had studied in Paris, told Joule about them. The two struck up a friendship and began to collaborate on sorting out what was right and what was wrong regarding heat engines.

Thomson thought, as Carnot had, that heat in equals heat out during the steam engine's cycle. After some back and forth, Joule convinced him this was wrong. They did recognize, however, the essential correctness of Carnot's insight that the work performed in a cycle divided by heat input depends only on the temperature of the source and that of the sink. When Joule's results were synthesized with Carnot's ideas, it became clear that a generic steam engine's efficiency—work output divided by heat input—differed from one (100 percent) by an amount that could be expressed either as heat out at the sink divided by heat in at the source, or, alternatively, as temperature of the sink divided by temperature of the source. Carnot's insight that the efficiency of the engine depends on the temperature difference was correct.

Of course, temperature has to be measured using the right scale. The correct one had been hinted at by Dalton and Gay-Lussac's experiments, in which true zero was –273 degrees Celsius. This meant that on this absolute scale, zero degrees Celsius should be taken to be 273 degrees and the boiling point of water as 373 degrees. A perfect cyclical heat engine with a source at 100 degrees Celsius and a sink at 7 degrees has an efficiency of 1 minus $^{280}/_{373}$, obtained by inserting in the engine efficiency formula a temperature of 280 (273 plus seven) for the temperature of the sink and 373 for that of the source.

There's only one way the efficiency can equal 100 percent—one way for the machine to be a perfect transformer of heat into mechanical energy: the sink has to be at absolute zero temperature. All forms of energy are, in principle, equal, but some are more equal than others. Mechanical energy can be transformed into thermal energy with 100 percent efficiency, but thermal energy can be transformed into mechanical energy only if you have a sink at absolute zero. Of course, since no temperature can be lower than absolute zero, the efficiency can never be greater than 100 percent.

In practical terms, the efficiency of a steam engine can be improved by increasing the temperature gap between the source and the sink. You might think the source is fixed at

373 degrees, the point at which water turns into steam, but pressure raises the boiling point of water. James Watt had already realized this. One of the improvements in his engines was the production of steam under pressure.

The unit of energy is now known as the joule, the heat developed in an electric wire is known as Joule heat, and the absolute scale of temperature, in which 0 degrees Celsius is 273 degrees, is known as the Kelvin scale. Why Kelvin? After a long life and a very distinguished career in science and public service, William Thomson saw his achievements marked in 1892 by becoming a lord, with the title Baron Kelvin.

Thermodynamics' Three Laws

The study of heat, its relation to other forms of energy, its means of transference, and its relation to temperature, was keenly investigated in the nineteenth century. The subject, broadly interpreted, was and still is known as thermodynamics. Many scientists of the time tried to define the axioms that lay at the base of their arguments, hoping that clearer formulations would resolve the ambiguities that continued to haunt the subject. Some of these debates seem archaic, but the struggle led to interesting new concepts. However complicated the arguments, thermodynamics' foundation has always been what are now known as its First and Second Laws, with a third somewhat obscure one occasionally thrown in for good measure.

In essence, the First Law of Thermodynamics says heat is just another form of energy and energy as a whole is conserved. The Second Law says you can't build a machine that will convert thermal energy into mechanical energy with 100 percent efficiency. Part of the reason the subject seems so complicated, at least from a historical point of view, is that the essential part of the Second Law was discovered by Carnot before the First Law was understood. Pedagogy, however, dictates they be called First and Second Laws, the logical scientific progression, even though their discovery proceeded in the opposite order.

Many scientists had thought about conservation of energy, but the experiments that unambiguously proved the First Law were performed by Joule and independently by a brilliant contemporary of his, the German Hermann von Helmholtz. Nevertheless, convention has them sharing the credit, for somewhat convoluted reasons, with Julius Mayer. As a young ship's doctor in the tropics, Mayer noticed that blood in the veins of patients there was redder than the blood in patients in northern climates. From there he started thinking about oxidation and eventually about the relation between energy generated in the body and heat produced by the body. A formulation of sorts of the First Law followed, couched in metaphysical terms that were often ridiculed even at the time.

I said that the struggle to understand and formulate thermodynamics' two laws led to some interesting concepts. None is more far-reaching than entropy, a notion introduced by Rudolf Clausius, another mid-nineteenth-century German scientist. Clausius tried to understand why mechanical energy is in some sense a "higher" form of energy than heat, and why it isn't possible to change heat into mechanical energy with 100 percent efficiency, although the opposite is true. In doing so, he managed to link together the degree of order and disorder in a system to the reversibility of a process. Thus, two boxes might contain the same amount of energy, but if it is in an orderly state in one and disorderly in the other, passage can only go from order to disorder; the process is not reversible. For Clausius, mechanical energy was more ordered.

An object rolling down a hill can come to a stop by friction, but the heat generated through that friction cannot be used to bring the object back to the top.

Clausius summarized his application of entropy to thermodynamics in two dramatic phrases that had a big impact at the time. They were (1) First Law: The energy of the universe is constant, and (2) Second Law: The entropy of the universe tends to a maximum. This second statement can also be paraphrased as "overall disorder always increases." Hermann Nernst followed up by enunciating what has come to be known as Nernst's Theorem, sometimes known as the Third Law of Thermodynamics. It states that an object's entropy goes to zero as its temperature goes to absolute zero. Absolute zero temperature is a state of complete order.

All three laws become clear if we return to Gay-Lussac's gas in a container and use the insights gained in the late 1850s through the study of molecular motion. In 1857, Clausius wrote an influential paper entitled "The Kind of Motion We Call Heat," relating average molecular motion to thermal quantities. Two years later, James Clerk Maxwell, probably the nineteenth century's most brilliant theoretical physicist, took up the same problem using a novel statistical approach. While an undergraduate at Cambridge University in 1855, Maxwell had shown that the rings of Saturn could not be either liquid or solid. Their stability meant they were made up of many small particles interacting with one another. In 1859 Maxwell applied this kind of statistical reasoning to the general analysis of molecules in a gas. He asked what sort of motion you would expect the molecules to have as they moved about inside their container, colliding with one another and with the walls. A reasonably sized vessel, under normal pressure and temperature, contains billions and billions of molecules. The speed of any single molecule is always changing because it is colliding all the time with other molecules. Thus the meaningful quantities are the molecular average speed and the distribution about the average.

Considering a vessel containing several different types of gas, Maxwell realized there's a sharp peak in the plot of the number of molecules versus their speeds. In other words, most of the molecules have speeds within a small range of some particular value. The average value of the speed varies from one kind of molecule to another, but the average value of the kinetic energy, one-half the molecular mass times the square of the speed, is almost exactly the same for all molecules. Now here's the jump: temperature is also the same for all gases in a vessel in thermal equilibrium. Let's go ahead and assume temperature is a measure of the average kinetic energy

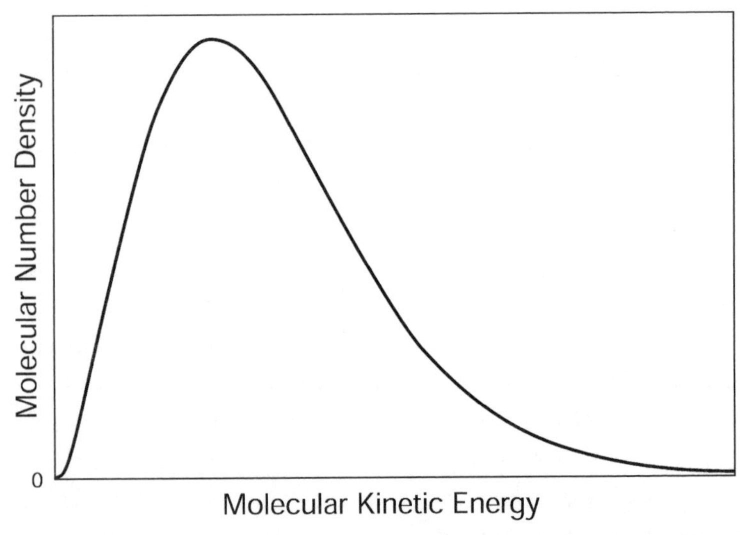

Plot of molecular number density versus kinetic energy for molecules in thermal equilibrium. The location of the peak shifts as temperature, in degrees Kelvin, changes.

of the molecules and then see how this helps to explain the puzzle of absolute zero's meaning.

Absolute zero is no longer a mystery. Kinetic energies are always positive. The average value of positive quantities is necessarily positive so the smallest value the average can take is zero and that value is reached only if all kinetic energies are zero. Zero is the average of a string of zeros. If we calibrate the gas by using the Celsius scale, the point at which every molecule is at rest, where every single one has zero kinetic energy, is –273 degrees. Absolute zero is absolute rest for all molecules. The connection of temperature with the average value of molecular kinetic energy works! In the end, it's not quite so simple: the attempt to bring every molecule to rest is one of the great challenges of modern physics. It brings up conceptual and practical problems involving all the subtleties of quantum mechanics, but that's a story for later in the book.

The meaning of the First Law of Thermodynamics is simple. The heat in the container is simply the sum of all the molecular kinetic energies. Thermal energy is just another way of describing motion energy, a summing of the very small mechanical kinetic energies of a very large number of molecules. The First Law of Thermodynamics is conservation of that energy; energy neither appears nor disappears.

According to Boyle's, Charles's, and Gay-Lussac's laws, molecules beating against the container walls cause pressure; the higher the temperature, the faster they move and the greater the pressure. Gay-Lussac's experiment, seemingly disproving the caloric theory, is also easy to understand: Divide a container, with vacuum on one side and gas on the other. Remove the divider and the molecules on one side spread over the whole container, but their average speed doesn't change. The temperature remains the same because temperature is the average molecular kinetic energy, not the concentration of caloric fluid. Thermodynamics aficionados might be puzzled remembering the Joule-Thomson effect in which a gas under high pressure cools its surroundings by escaping through a nozzle into a lower-pressure environment. In this case the expanding gas does work and loses energy, thereby lowering its temperature and drawing heat from its immediate neighborhood. By contrast, during expansion into an adjacent vacuum, no energy is lost and temperature is unchanged.

Entropy and Life

Shortly after Maxwell's successful analysis of molecular motion, Vienna's Ludwig Boltzmann began to think about how that analysis could be used to understand disorder and reversibility. After many years, he managed to give a statistical interpretation of Clausius's notion of entropy and to explain in all its depths the Second Law of Thermodynamics.

The notion that heat flows from hot to cold and not the other way around could be phrased in terms of molecular motions. You know what will happen to molecules in a container when you wait a while, allowing them to collide with one another: the faster ones slow down and the slower ones speed up, not the other way around. This means the hotter part becomes colder and the colder part becomes hotter: thermal equilibrium is reached. If you put an ice cube in one corner of a large box at room temperature, it melts. If you put hot water in another corner, it doesn't form an ice cube.

Boltzmann was a sensitive, moody man, subject to fits of deep depression. After his death by suicide in 1906, his native Vienna erected a monument to him. Boltzmann, who loved music, would have been pleased to know he was buried in the same part of the large Zentralfriedhof Cemetery as

Beethoven, Schubert, Brahms, and Strauss. But there are no musical notes on his tomb. Rather, engraved on the bas-relief above his head is Boltzmann's formula for entropy, $S = k\log W$. In it, S is entropy, k is a constant now known as Boltzmann's constant, and W is a measure of the number of states available to the system whose entropy is being measured.

The concept of entropy has by now entered our lives in all sorts of ways outside of physics experiments. In throwing dice, you are more likely to roll a seven than a three because seven can be made by six and one, five and two, or four and three, while three needs only a two and a one. Seven has greater "entropy"—more states, more ways to roll it. Going back to Gay-Lussac's container full of gas, remember how the gas spreads out when the divider is removed. Entropy has increased because there are more different possible molecular trajectories. The chance of seeing the gas retreat back to one side of the container after it has spread goes to zero very rapidly as the number of molecules increases. Such a retreat corresponds to decreasing entropy, more restricted motions for the molecules. Just as there are more ways to roll seven than three with two dice, there are more ways to fill a whole container rather than half a container with a fixed number of molecules.

Can we put the gas molecules back on one side of the container? The simplest way to do that is to place a piston on the right side of the container, applying a gentle pressure, and squeezing the molecules to the left until they are back in their original configuration. There's a subtlety, however. The compression heats the gas, so the molecules are really back in their original configuration only after that extra heat is removed. If that heat could then be used to drive the piston in a 100 percent efficiency conversion, I could construct a perpetual motion machine, expanding and contracting the gas in the container. But that's impossible; the conversion of heat to work is never 100 percent efficient. The entropy of the gas decreases as I push it back to one side of the container, but the overall entropy of the piston, the cooling system, and the container increases.

The connection of probability to information begun by Boltzmann is by now the cornerstone of fields far from thermodynamics, such as modern information and communication theory. It also raises some interesting questions. If disorder grows, if we follow Clausius's dictum that the universe's entropy always increases, how is it possible to have the orderly progress of life, the transmittal of genetic information, and the steady replication of that information? The answer is subtle.

A living organism maintains its extraordinary degree of order by the continual alteration of energy from nutrients into mechanical energy and heat. After death, the metabolic changes cease and the organism's disorder and hence entropy rapidly increases. Nevertheless, the assembly and maintenance of a live creature involves the continual shift of nutrients from a more disordered to a more ordered state, one with lesser entropy. Thus, while the entropy of a living individual decreases, the entropy of the individual *plus* that of the other individuals, plants, animals, oceans, and earth that surround him or her increases continually. Life is possible, but Clausius was right: the entropy of the universe increases.

Viewed correctly, life is compatible with the basic principles of physics and chemistry. More than that, the new late-nineteenth-century ways of thinking about biological systems fit naturally into this scheme, this way of looking for patterns in large numbers, in assigning precise properties to classes, to groups, and not to individuals. The great French molecular biologist François Jacob has convincingly argued that new insights into the nature of heat, the statistical understanding of thermal motion, and the overarching principle of energy conservation changed the way scientists think about life: "At the beginning of the nineteenth century, an organism expended vital force in order to perform its work of synthesis and morphogenesis; at the end of the century, it consumed energy."

By 1900 scientists didn't talk any more about either vital forces or caloric fluids. Energy was the key concept—its uses, its organization, and its many manifestations. Chromosomes and the duplication of cells were studied under the microscope and genetics was poised to emerge as a field. The seemingly contradictory estimates of the lifetime of the Earth necessary for evolution and the lifetime allowed by the heating power of the Sun would soon be solved.

Science was taking a series of dramatic turns in dealing with new insights. A picture of a much older Earth that had gone through many warm and glacial periods was beginning to form, a picture that needed to join smoothly to the insights of evolution. All of the sciences were involved in unraveling these new puzzles. After looking backward, scientists now turned their glances forward and began to conjecture about the near and distant future of the Earth, how hot or cold it would become, and what that might mean for life.

Chapter 3

READING THE EARTH

IN LATE JULY 2000, the Russian icebreaker *Yamal* left its home port of Spitsbergen, Norway, headed toward the North Pole. The ship had a few scientists on board as well as a group of tourists intent on both seeing the Arctic and, in a symbolic move, standing on the Pole. The *Yamal* was equipped to crunch through ten-foot-thick layers of ice, but, to the surprise of everyone aboard, the ship found only open waters to the north, punctuated intermittently by thin layers of ice. When the Global Positioning System's navigation device told the captain he had reached the Pole, the ship was surrounded by water. Under a clear sky, ivory gulls flew overhead, the first time these birds had been spotted that far north. The *Yamal* had to travel an additional six miles to find ice firm enough to allow the hundred or so passengers to get off the boat and take their symbolic stand.

The ship's captain, who had made the trip to the Pole many times, said he had never seen open water there. The oceanographer James McCarthy, director of the Museum of Comparative Zoology at Harvard University and a lecturer on the *Yamal*'s trip, could only express his surprise and consternation. He recalled an earlier journey he had made to the Pole that required breaking through thick sheets of ice.

During World War II, the Royal Canadian Mounted Police tried to take one of their ships, the *St. Roch*, through the

Northwest Passage, which connects western Canada, just below its Alaskan boundary, with the waters of the Atlantic. This never was an easy trip. Many nineteenth-century explorers met their deaths in frozen campsites, watching their trapped boats crushed by ice packs. The *St. Roch* had been frozen in through two winters, moved with great difficulty in the summer, and only reached the Atlantic twenty-seven months after it set out. In the summer of 2000, however, a new ship, the *St. Roch II,* completed the 10,000-mile trip in under a month, including several stops along the way. The captain of the *St. Roch II* described his trip: "There were some bergs, but nothing to cause any anxiety. We saw some ribbons of multiyear ice floes, all small and fragmented and were able to steer around them."

The warming of the Arctic regions has been dramatic. The Columbia Glacier, which ten years ago provided an imposing sight in Prince William Sound to cruise ships sidling up to its 200-foot-high wall, has since retreated more than sixteen miles, leaving in its wake a stretch of open land. The once permanently frozen ground of much of Alaska has begun to thaw. As it does, softened roadbeds buckle. Thriving beetle populations are attacking forests of white spruce. Summer temperatures in Fairbanks now reach the 80s for weeks on end. Scientists say the average temperatures have gone up in a short time by at least 5 degrees in Alaska, northern Canada, and Siberia, and by as much as 10 degrees in parts of these Arctic regions. *Ursus maritimus,* the great polar bear and North America's largest land carnivore, often called the "Lord of the Arctic," is on the average 10 percent thinner than he was twenty years ago. The bears' already short hunting season has been decreased by three weeks.

The first report of the *Yamal* meeting open water at the Pole made the front page of the *New York Times*. Follow-up stories on later days, no longer on the front page, included cautionary remarks, but there is little doubt that the Arctic is getting warmer. As Dr. McCarthy, the lecturer on the *Yamal,* put it, "What was really unusual was that over a period of two weeks we never had a day of what would be considered normal ice. When we reached the pole and found open water, that simply punctuated what we were seeing everywhere. These were conditions that did not seem representative of a transient phenomenon."

The melting of Arctic ice is worrisome, but just how worrisome should it really be? How much are local and global temperature shifts under our control, how do humans contribute to them, and how much are they simply part of natural fluctuations? What is the ripple effect, how reversible are changes in climate, and what kinds of time lags are involved? How do we go about changing our patterns of resource usage and who makes the decisions? These are all complex questions, among the most difficult we are faced with as inhabitants of the globe. The answers are not clear; all we can do is make some estimates of what is likely to occur as we follow each one of a number of paths into the future.

Let's start with something that's comparatively easy to determine, namely, how much the *Yamal*'s observations are anecdotal and how much they reflect a real change in weather at the Pole. Arctic sea ice thickness, as determined by submarines navigating in polar regions, measures only 60 percent of what it was a few decades ago. In the 1990s, the United States Navy invited scientists aboard their Sturgeon-class submarines for five cruises through Arctic waters, allowing them to analyze the echoes of sonar beams bounced off the bottom of the ice sheets. Data from the 1980s is still classified, but Andrew Rothrock, Yangling Yu, and Gary Maykut from the University of Washington have found publicly available data sets of sonar echoes from submarine cruises in 1958 and 1976. They checked them against their results from the 1990s. Where comparisons could be made, the researchers found the depth of ice below sea level in the Arctic mass decreased from an average of ten feet to one of six feet. These observations match those obtained bouncing radar signals off the ice from a low-flying satellite. The warming climate shift we are all experiencing appears to have hit the polar regions hardest for reasons that are not altogether explicable. But it is real.

Is the very significant Arctic warming a local fluctuation or is it the portent of a growing trend over the Earth? To compare, let's look at changes in the Antarctic. There is less evidence in the Antarctic of dramatic warming, but the stakes are higher because warming in the Antarctic has a greater potential for affecting us all, even those far from the region. The danger is related to those very same open waters sighted by the passengers on the *Yamal*. The North Pole is covered by water, usually with a layer of ice on it, but the Antarctic is a land mass covered by ice.

The melting of already floating Arctic ice doesn't displace any additional water, but if Antarctic ice melts, slides off land, and enters the oceans, water levels rise around the world. The most vulnerable of all such large ice masses is the West Antarctic Ice Sheet (WAIS), perched at the southern end of the Earth, near where the Atlantic and the Pacific meet. The big sheet already partially floats in water, creating the Ross Ice Shelf, which bottles up the larger mass of ice on the continent. If that block, estimated to contain a million cubic miles of ice, were to melt, it would lead to global flooding, disappearance of most ports around the world, and an average rise in the oceans of some fifteen feet. Bangladesh would be underwater, the Netherlands would be endangered, and much of Florida and Louisiana would be gone.

The difficulties in assessing how likely it is for the block of ice to move led the Intergovernmental Panel on Climate Change, a multinational body, to conclude in 1995 that "Our ignorance of the specific circumstances under which West Antarctica might collapse limits the ability to quantify the risk of such an event occurring, either in total or in part, in the next hundred to one thousand years." The West Antarctic Ice Sheet wouldn't even need to melt to unleash this catastrophe; by simply sliding off land, it would displace the ocean with the now-submerged volume of the ice. It is particularly frightening to think that the chain of events that might cause this disaster has already been set in motion. The total disappearance of the great ice sheet is a worst-case scenario, but it is one of many risks we need to evaluate. This is hard because the risks are intertwined. A rise in temperature at one place can trigger climatic conditions that will set off an environmental disaster elsewhere; a rise in air temperature can have a big impact in changing ocean levels.

Global warming, or, for that matter, global cooling, is obviously a complicated problem, one that is best dealt with by looking at all the agents of change. There are four main contributors to the temperature of the Earth's surface. The first is the Sun, our principal source of heat. When the Earth is tilted toward the Sun, it's summer, and when tilted away, it's winter, but there's more to that story because the Earth wobbles as it rotates on its elliptical path around our Sun. The second contributor is the heat generated directly in the Earth. This heat, most visibly displayed in the eruption of volcanoes and more subtly in the rise in temperature as one descends into a deep mine, is largely produced by the radioactive decay of elements in the outer parts of the Earth and by heat from the core, stored there in the Earth's early formation. Though some of its manifestations are spectacular, the amount of heat generated in the Earth is still minuscule compared to the amount that sunlight provides to the Earth.

Third, oceans are big contributors to global temperature shifts, largely through the round-the-world motion of giant masses of water in what looks like a conveyor belt system. Many of these currents, the Gulf Stream being a good example, have been studied for hundreds of years, but others, like El Niño, which is more of an upwelling than a current, are relatively new to us. The fourth and final major contributor to global temperature is the blanketing of the Earth in its atmosphere, a cover that both shields the Earth from harmful radiation and keeps much of the terrestrial warmth generated by the Sun from escaping into space.

There are other contributors to climate change. Collisions with asteroids are particularly dramatic. In that case, the main source of climate change is not the collision itself, a local phenomenon, but the dust injected into the atmosphere following the impact.

In other words, Sun, Earth, water, and air explain Earth's temperature. The problem of understanding and then predicting climate changes resulting from alterations in the four is hard because all four contribute to climate simultaneously. Through the use of supercomputers and increasingly sophisticated modeling, we are entering an era when predictions are becoming more reliable, but accurate long-term forecasts are still very much a work in progress.

Our planet's evolution has played itself out against the backdrop of the heat and the cold determined by our environment and these four initiators and perpetuators of climate change. The story of this past is one of great and unexpected changes, of stops and starts, abrupt shifts and reversals. It has been uncovered by painstaking scientific work, by improvements in measuring devices, and by human ingenuity. How we effectively use this knowledge is the part of the story that has yet to be written. The motion of the Earth relative to the Sun is the greatest factor in determining climate and the only factor that is completely out of our control. That makes it a good place to start our story of the past.

Copernicus's Harmony

The rotation of the Earth about its axis every twenty-four hours gives us the change from day to night as we face away or toward the Sun. The inclination of the Earth's axis of rotation with respect to the plane of its orbit changes very little over the course of a year. When the Northern Hemisphere tilts toward the Sun, it is summer. Six months later winter descends on the Northern Hemisphere as it tilts away from the Sun. That's the first explanation—and probably the most important—of why it's hot and why it's cold. As always, the devil lies in the details—the wobbles and subtle changes of that orbit.

In 1543 Copernicus published *De Revolutionibus Orbium Caelestium,* a book in which he made the case that the Earth revolves around the Sun, not the other way around. Others, as far back as Aristarchus in the third century B.C.E., advocated the same view, so the real change, as the historian of science Thomas Kuhn has emphasized, is that Copernicus was the first to place that motion in a coherent mathematical scheme and the first to realize that the Earth's motion might provide the explanation to a series of puzzling observations. Advancing such an idea was so heretical that Copernicus, a cautious rebel, made sure his book was not released until he was on his deathbed.

Even after deciding the Earth revolved around the Sun rather than vice versa, Copernicus clung to Greek ideals of perfection; the orbit had to be circular. But by Kepler's time, in the early seventeenth century, a new realism was in place. The Renaissance in science, as in art, was marked by a relatively faithful rendition of the outside world, complete with shadowing and perspective. Copernicus's elegant picture was replaced by Kepler's painstakingly accurate measurements demonstrating that planets move around the Sun in elliptical, not circular, orbits.

Once Kepler and his followers showed the orbits were ellipses, the notion of perfect shapes for celestial bodies was abandoned. In his 1687 compendium, the *Principia,* Newton not only gave a simple explanation of Kepler's laws, thereby setting a style for science that continues to this day, but went on to calculate the equilibrium shape of a rotating Earth. There was no longer any reason for the Earth to be thought of as a sphere and, according to Newton, it wasn't one. The Earth is flattened at the poles and bulges at the equator. Newton even calculated the ratio of the distance between the poles to that across the Earth at the equator. He found it to be approximately 230:231, remarkably close to the best present-day measurements. This calculation explained a well-known puzzle. In the 1670s Jean Richer had been sent to South America by the Paris Academy of Sciences to perform astronomical measurements. In Cayenne, near the equator, Richer observed that his pendulum clock, set in Paris, seemed to be about 2.5 minutes a day slower. The period with which a pendulum swings had been shown to be unchanging by Galileo, but now there seemed to be a minute correction. Since the value of the period depends on gravitational force, and if the Earth is not a sphere, the gravitational force on the pendulum's bob is not the same in Paris and Cayenne. A 230:231 ratio of the axes translates into a 2.5-minute-a-day difference in the pendulum clock!

Newton also realized that the equatorial bulge would lead to small changes in the alignment of the Earth's rotation axis with respect to the plane of its orbit about the Sun. The gravitational pull from both the Sun and the Moon on a nonspherical Earth means the axis's orientation will very slowly, perhaps over thousands of years, trace out a circle in the sky, all the while keeping a fixed angle with respect to the plane of the orbit. The computation of the period of that circular motion, known as precession, was too much for even Newton. That calculation was finally done in 1754 by the great French mathematician Jean Le Rond d'Alembert. The answer is 22,000 years.

The precession of the Earth's rotation axis, though much slower, bears a resemblance in its appearance to a rapidly spinning top. If the top is set spinning with its axis exactly perpendicular to the ground, the axis remains perpendicular. If, as is usually the case, the axis is at an angle to the ground, it slowly describes a circle around the vertical as the top wanders across the floor. In addition, the top can bob up and back a little bit, an additional kind of motion known as nutation. The Earth's rotation axis has the same feature, again completing this motion very slowly

Precession explains an old puzzle—the shift of the location in the night sky around which stars appear to rotate. The fixed star is now Polaris, the North Star, but in 3000 B.C.E. it was Thuban in the Draco constellation. The apparent rotation of the stars about Polaris is caused by the real rotation of the Earth about its axis, an axis that now points at Polaris, but that has been moving slowly over time. The period of that precession, as calculated by d'Alembert, is many thousands of years.

Even though calculation of the periodicities of the Earth's precession were not attempted until 250 years ago, extraordi-

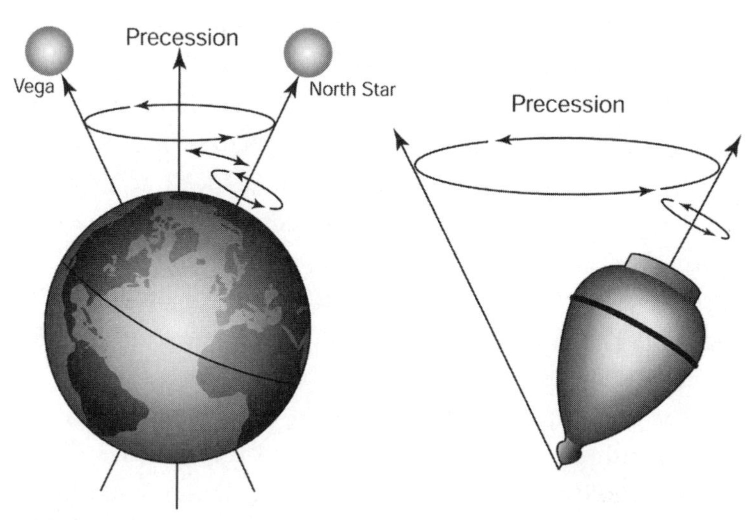

How the precession of Earth's axis of rotation is like the precession of a top

narily accurate measurements of them were probably made 4500 years ago. This point was driven home to me on seeing a photograph of the Egyptian pyramids on the cover of the November 16, 2000, issue of *Nature*. The article in the issue attempts to resolve a puzzle that has been known for over a hundred years concerning the eastern and western sides of the great Khufu (Cheops) pyramid, which are aligned with true north to within a twentieth of a degree, an accuracy comparable to the best attained by any viewer of the sky before the telescope was invented. This extraordinary precision strongly suggests stellar observations were a key to the orientation. The puzzle is that the alignments of the eight pyramids built between 2600 and 2300 B.C.E. have an accuracy that varies according to when the pyramids were built.

In the issue of *Nature* I cited, an English scholar named Kate Spence suggests a solution to the puzzle. Because of the precession, true north—the point about which stars rotate—

was not Polaris when the pyramids were built. There also was no other bright star at true north, so Spence thinks the Egyptians set the direction as best they could by drawing an imaginary chord between two bright stars, Mizar and Kochab, which then bracketed north's location. Spence's argument is that the precession of the Earth changed the orientation of the chord. The most reliable alignment of the pyramids to true north happened to occur during the building of the Khufu pyramid. The older pyramids are aligned slightly more in one direction and the newer ones in the other.

The data fit. In addition, the movement of the stars is known with greater accuracy than any other records, so the argument allows us to date the building of the pyramids to within a few years, an improvement over estimates made by written records or carbon dating. Of course the argument presupposes that each pharaoh had a measurement of north made in his reign and built his pyramid accordingly. That doesn't seem unlikely. The argument would be clinched if archaeologists found a record of the ceremony of setting true north.

In 1842, almost a hundred years after d'Alembert's calculation, a French mathematician named Joseph Alphonse Adhemar followed up earlier calculations on variations in the Earth's orbit around the Sun. In a book entitled *Revolutions of the Sea,* Adhemar proposed that the 22,000-year precession of the Earth's axis would lead to a 22,000-year cycle of ice ages. According to Adhemar, the key to causing such ice ages is a particularly cold winter in which a hemisphere is maximally far from the Sun in its elliptical orbit, and maximally tilted away from the Sun. This confluence occurs every 22,000 years in the Northern Hemisphere and, of course, every 22,000 years in the Southern Hemisphere.

Adhemar was wrong. The smaller amount of heat in the

winter is exactly balanced by a larger amount of heat in the summer. In other words, a particularly cold winter is followed by a particularly hot summer, leaving the yearlong amount of heat unchanged. Yet Adhemar was prophetic in suggesting a link of planetary motion to ice ages and in doing so at a time when ice ages were only beginning to be discussed.

The emergence of these discussions was possible because probing questions about the Earth's past came up between d'Alembert's 1754 calculation of the precession and Adhemar's 1842 book. In 1654 Bishop Ussher said his biblical studies led him to conclude that the Creation occurred in late October of 4004 B.C.E., a date not that different from the earlier estimates of experts ranging from St. Augustine to Kepler. This kind of dating was seriously questioned at the end of the eighteenth century. What if the age of the Earth were to be determined by scientific observation rather than religious narrative? At times flirting with heresy, the role of observation and measurement was gaining credibility and acceptance.

Georges-Louis Leclerc, Comte de Buffon, was the eighteenth century's best-known and most popular writer about science. His thirty-six-volume *Histoire Naturelle* was accessible to the lay public and became a great, albeit controversial, best-seller. Buffon thought the Earth originated as a sphere swept off the Sun by a collision with a large comet. Newton had calculated that a red-hot globe of iron the size of the Earth would take 50,000 years or more to cool. Buffon tested this idea by measuring the cooling of small globes cast in a foundry. With his new calculation in hand, he came up with an Earth lifetime of 80,000 years. Buffon then neatly divided this into seven epochs of successive cooling, ranging from white-hot to the present. By doing so, Buffon sought to reconcile his version of genesis with the biblical story, each day of creation an epoch. There are many problems with Buffon's idea. Newton, for one, never believed in a cometary origin of the Earth, but the idea represents the ferment of ideas of the period.

Puzzles were cropping up as 1800 approached. Fossils of extinct animals were uncovered and surveyors of the new canals were commenting on rock strata that seemed to lie in the same sequence, suggesting a geologic past. Perhaps the Earth really was very old and had a variety of climates in its past. In 1795 the Scottish chemist James Hutton went so far as to propose that the Earth had no beginning in time and would have no end.

The Minister, the Lawyer, and the Fossil Fish Expert

In 1840 three unlikely companions came to agree on the existence of past ice ages. The oldest of the three was the Reverend William Buckland; born in 1784, he became a minister in 1809. He pursued his interests in mineralogy and the new subject of geology with such fervor that by 1818 he was a member of the Royal Society and by 1819 he was Oxford University's first professor of geology. Buckland sought to reconcile science and theology, becoming the leading exponent of Catastrophism, a school of thought predicated on the notion that geological shifts are caused by disasters, cataclysms, and floods. The first and greatest of them was, of course, Noah's Flood, or, to borrow part of the title of Buckland's most famous work, "the Action of a Universal Deluge."

The second of the trio, Charles Lyell, was thirteen years younger than Buckland. He entered Oxford University to study law and eventually became a barrister, but Buckland's lectures captured his attention and slowly turned him in the direction of scientific research. In 1824 the forty-year-old

Buckland and his twenty-seven-year-old pupil took a geological study tour through Lyell's native Scotland. The trip strengthened Lyell's resolve to continue what had been an avocation; by 1826 he was also a member of the Royal Society. In 1827 he abandoned the legal profession in order to devote himself full time to science.

Though Buckland and Lyell remained good friends, their views diverged. Lyell became the principal spokesman of Uniformitarianism. This school of thought believed erosion gradually whittles mountains down and successive earthquakes slowly elevate the crust; it also believed that the survival and extinction of species is caused by the ebb and flow of more and less favorable climatic conditions. Lyell avoided, when possible, theological controversies, simply saying the evidence should speak for itself. Not surprisingly, Buckland's influence waned as the years went on while Lyell's grew, enhanced in great measure by his three-volume *Principles of Geology.* Charles Darwin took the newly published first volume with him on the voyage of the *Beagle,* making sure the second volume reached him in Montevideo. These books became the nineteenth century's invaluable geology texts and Lyell became the younger Darwin's closest friend. As Darwin said in his autobiography, "The Science of Geology is enormously indebted to Lyell—more so, as I believe, than to any other man who ever lived."

Lyell's followers, the Uniformitarians, struggled against Buckland's, the Catastrophists, to define the origins of the Earth, but there were puzzles neither could explain to their own satisfaction. One of these was the presence in the countryside of large rocks, usually referred to as erratic boulders. These boulders, often as large as a house, were different in composition from anything in their vicinity and had obviously been transported to their locations by some unknown force. The Catastrophists claimed giant tidal waves moved the boulders, while the Uniformitarians proposed that the huge rocks had been embedded in slowly drifting icebergs that carried them from the far north to their present location. Both models were consistent with the notion of a great flood, so there was no theological opposition, but neither model could explain satisfactorily the presence of boulders and other deposits at high altitudes in mountainous regions. Had the flood engulfed the whole Earth?

As Uniformitarians and Catastrophists debated their conclusions, the findings of the third member of the trio emerged. Ten years younger than Lyell, twenty-three years younger than Buckland, Swiss-born Louis Agassiz took degrees in philosophy and medicine while pursuing his interests in botany and paleontology. In his early twenties he laid the plans for his five-volume study of fossil fishes, and by twenty-five he was a university professor of natural history in Neuchâtel. The work on fishes made him a young scientific star and by 1838 a foreign member of the Royal Society.

Young Agassiz, the fossil fish expert, conjectured that the boulder puzzle was the natural outcome of an earlier, much colder, age during which Europe and large parts of North America lay under a sheet of ice. As Agassiz saw it, stones with deep scratches, moraines, and, of course, the erratic boulders were all caused by advancing and receding ice. Agassiz also claimed that the glaciers in his native Switzerland were simply scaled-down versions of those great ice fields, the difference being one of degree, not one of kind. The idea that glaciation causes erratic boulder placement did not originate with Agassiz, but he was its driving force. His energy, his dynamism, the wide sweep of his ideas, and his clout in the scientific world all combined to make him impossible to ignore.

Agassiz's interest in glaciers began in 1836 when he went

The Reverend Buckland outfitted as a "glaciologist"

to spend the summer with his old friend Jean de Charpentier, an amateur scientist. As is often the case, the visit was part work and part play. There were many fossils near Charpentier's house that Agassiz wanted to study, but, as the visit wore on, the focus of discussions turned increasingly to observations of glacier movement. Charpentier believed that the glaciers had once filled the nearby valleys; at first Agassiz was skeptical, but the more he examined the facts, the more convinced he became. The two friends took long walks and then went to study the nearby Chamonix glaciers to see with their own eyes that the phenomena were not unique to Charpentier's valleys. Agassiz even had a hut constructed for himself on the edge of the Aar glacier so he could spend day and night without distractions examining the movement of the ice. Once he was convinced of ice movement, he made the case to the public; by 1840 the prolific Agassiz had written a two-volume treatise on glaciers. He spoke with authority the language of science, he could back up his observations with geological data, and he painted an exciting picture of a past with glaciers extending down to the Mediterranean.

In the summer of 1838 the Reverend and Mrs. Buckland, on a tour of continental Europe, paid a visit to their distinguished young Swiss colleague. However, a trip to the Alps with Agassiz was not enough to shake Buckland's belief in tidal wave movement of boulders. Agassiz and Buckland met again in the summer of 1840 when Agassiz went to Britain to study fossil fishes. At this time he was invited by Buckland to join him and the distinguished geologist Roderick Murchison on a study trip through northern England and Scotland. During the trip, the three of them examined the terrain, the grooves in the land, the small rocks, and the boulders. In seeing with fresh eyes all this evidence, Buckland came to believe that glacier movement had to be the explanation. Once persuaded, Buckland became a proselytizer of the doctrine. He contacted his old friend and disciple Lyell, who recognized the power of the idea and its essential correctness. Slow glacier movement was more in line with Uniformitarianism than with Catastrophism anyway, so if Buckland could be convinced, Lyell was not going to lag behind. By the late fall of 1840, the minister, the lawyer, and the fossil fish expert, now united in their beliefs, read papers at the Geological Society on the evidence of past existence of glaciers in England and Scotland.

Buckland continued to perform valuable public and scientific services. He died in 1856, three years before the publication of Darwin's *Origin of Species,* a book he surely would not have agreed with. He is buried in the yard of his church at Islip, near Oxford. Lyell's fame continued to grow. He was knighted in 1848 and made a baronet in 1864. He became Darwin's mentor; though initially resistant, he eventually embraced the concept of natural selection. Lyell died in 1875 and was buried near Newton in Westminster Abbey, the pantheon of Britain's greats.

Agassiz moved to the United States and became a professor at Harvard, one of the first of a long line of distinguished European scientists immigrating to the United States. Despite Agassiz's adventurous spirit and willingness to accept bold new ideas, he never embraced the concept of evolution, continuing until his death to affirm the alternative notion of independent creation. He died in Cambridge, Massachusetts, in 1873 and is buried in the same Mount Auburn Cemetery where my mother and my wife's parents lie. The monument on Agassiz's grave, shaded by pine trees brought from his native Neuchâtel, is very different from theirs, however. It's a boulder from the glacier of the Aar where his simple hut once stood as he watched the glaciers move.

Cycles of Ice

One might think that once the notions of moving glaciers and past ice ages had been acknowledged by the leading proponents of both Uniformitarianism and Catastrophism, there wouldn't be any further opposition. However, universal acceptance didn't occur until the early 1870s, and even then the cause of ice ages remained a mystery.

As I said earlier, Adhemar's 1842 suggestion that the Earth's precession might lead to ice ages was quickly shown to be wrong since the exceptionally cold winters Adhemar predicted would be followed by exceptionally hot summers, leading to unchanged yearly heat balances. His idea was nevertheless taken up twenty years later by a self-taught Scot named James Croll. Croll was born poor in a village in Perthshire in January 1821; his family's circumstances did not allow for much schooling, but Croll continued to read and study while working successively as a wheelwright and a carpenter. An old elbow injury that never properly healed made those occupations increasingly difficult, so Croll turned to more sedentary labors. He worked in a tea shop, sold electrical devices to alleviate body pains, ran a small hotel, and sold insurance. He didn't do well at any of these and seemed destined to fail. In 1859 his wife's illness forced Croll to move with her to Glasgow where her sisters could nurse her. As Croll later said, a menial position at the Glasgow Andersonian Museum, which he took up in 1861, was his salvation: "My salary was small, it is true, little more than sufficient to enable me to subsist; but this was more than compensated by advantages for me of another kind."

The museum had a good scientific library and Croll had time on his hands. Enthralled by the ice age problem, Croll made a very interesting observation. He asked himself what might happen if there were a series of particularly cold winters. He thought lower temperatures would mean more snow and therefore more sunlight reflected back. In other words, more snow would make the Earth cooler. That would mean even more snow would fall. Croll was discovering how small differences in the temperature are enhanced through what we now call a feedback mechanism. In this case, slight orbital differences might trigger an amplified cooling of the Earth, which would not necessarily be balanced by summer warmings.

The first feedback mechanism Croll considered was

snowfall; ocean currents came next. The standard sailing route to America from northern Europe was south past Portugal, Madeira, and the Canaries, and then west across the Atlantic following the east-west trade winds. After loading cargo in the Caribbean, a ship might head north along the Gulf Stream and then sail east back to Europe, completing a great sweep. Croll thought the currents flowed this way at least in part because the coast of Brazil juts out into the Atlantic like a giant triangular wedge; the westward drift of warm equatorial water hits the upper side of the wedge and is deflected northward. Croll reasoned that if the currents were a little different, the warm westward-moving water would hit the lower side of the Brazilian wedge and be directed south instead of north. An increase of the ice sheets in one hemisphere might be just enough to change the winds and alter the side of the wedge hit by the westerly current. This would lead to an even colder Northern Hemisphere, one covered by an ice sheet. Croll now had a second example of a feedback mechanism. At this point Croll was confident that small climatic changes would, by feedback amplification, periodically turn cold weather into ice ages.

Croll also had the advantage of recent astronomical calculations. Adhemar's predictions were based on d'Alembert's calculation of the effect on the Earth's orbit of the Moon, the Sun, and the Earth's equatorial bulge. In the thirty years between Adhemar's book and Croll's calculations, a French mathematician named Urbain-Jean-Joseph Le Verrier computed the perturbations of the Earth's orbit due to the other planets in the solar system. According to Le Verrier, the eccentricity (roughly, shape) of the Earth's elliptical orbit varied with a period of 100,000 years. He also calculated the period of the tilt of the orientation of the Earth's rotation axis with respect to the plane of its orbit about the Sun.

Putting all this information together, Croll argued that high eccentricity eras, lasting about 100,000 years, would cause alternating ice ages every 11,000 years in the two hemispheres, while times of low eccentricity would have no ice ages. He also predicted that the last set of ice ages had ended some 80,000 years ago, leaving us in an interglacial epoch. Croll's results, first published in 1875, were quickly accepted. Within a year of his book's appearance, the self-taught Scot was a Fellow of the Royal Society and an honorary graduate of St. Andrews University. His glory was only short-lived; the tide was turning against him when he died in 1890. Measurements made toward the end of the nineteenth century indicated the last ice age had ended more recently, not the 80,000 years ago that Croll claimed. By 1911, when the *Encyclopaedia Britannica*'s eleventh edition was published, his entry read:

> The soundness of Croll's astronomical theory regarding glacial period has been criticized by E. P. Culverwell in *Philosophical Magazine* and by others; and it is now generally abandoned. Nevertheless it must be admitted that his character as a scientific worker under great discouragements was nothing less than heroic.

It's ironic that the first article reestablishing the essential correctness of Croll's approach was written in 1914, when the *Britannica*'s ink was barely dry. The author of the 1914 paper was a thirty-six-year-old Serbian mathematics professor. Milutin Milankovitch, working with better data than Croll had, believed he could determine the climate of the past and predict future ice ages. It took him ten years before he got his first break. In the early 1920s a German mathematician named Ludwig Pilgrim had taken Le Verrier's calculations a step further by determining the period of the variation of the Earth's tilt. This turned out to be 41,000 years. The ice age theory now involved not two but three separate periods: precession (22,000 years), eccentricity (100,000 years), and tilt (41,000 years). Milankovitch redoubled his systematic attack on the climate problem: the amount of solar radiation that hits the Earth at any given spot, at any given time, depends on only two factors, the distance from the Sun and the tilt of the Earth; now, thanks to Pilgrim, he had both sets of data.

By 1930 Milankovitch had calculated the variation in solar radiation as a function of geological time at eight separate latitudes ranging from 5 to 75 degrees north. His first results matched well with the history of Alpine glaciers, so he was optimistic, but he still needed a good estimate of feedback effects. By 1941, Milankovitch was able to provide a complete thermal history of the Earth over more than a half million years. The temperature minima, which followed from the interplay of the eccentricity, tilt, and precession periods, had no simple periodicity of their own. His plots showed a long interglacial epoch between 200,000 and 400,000 years ago, but they also showed nine sharp minima in the past 650,000 years—nine significant ice ages.

During those ice ages, the summers were about 12 degrees Fahrenheit colder than today's, cold enough to cover the Earth with great ice sheets. Milankovitch's model met with considerable success. However, by the mid-1950s, the new techniques suggested dates at variance with his results. As so often happens in science, a period of confusion and conflicting data came next. By the mid-1970s, James Hays, John Imbrie, and Nicholas Shackleton convincingly proved the model was correct. The basic line of thinking, which stretches back 125 years to Croll's work, is now accepted. Ice ages follow what all geology textbooks call Milankovitch cycles.

The Tundra's Bloom

Ice is one of the best places to look for a record of warming and cooling, for evidence of Milankovitch cycles, and for sudden shifts in temperature. Ice that has accumulated for very long periods is particularly good. The best deep ice is in Greenland and Antarctica. Scientists look for spots where snow has been piling up without melting for thousands of years, slowly pressing the underlying layers into even sheets of ice. They drill deep into that ice, pumping fluid to equalize the ice pressure and keep the hole from collapsing. At a predetermined depth, they stop the drill, scoop out an ice sample, and lift it back to the surface. In Greenland, drills have gone down two miles, all the way to bedrock, and come back carrying a history of 100,000 years of snowfalls.

Ice layers are like the rings in trees, a journal of past climates. The thickness of a layer in an ice sample tells how much snow fell that year. Coarser, heavier dust particles mean strong winds; volcanic ash is the sign of an eruption. By 200 feet down the pressure is enough to trap air bubbles, bubbles that can be analyzed for levels of methane and carbon dioxide. Bacteria in tropical swamps produce methane; a methane increase means more tropical wetlands, i.e., warmer climates. These are some of the indirect methods of finding the temperature of a given year. There are also direct ways to do it.

Not all ice looks alike and the variations—some detectable only with sophisticated instruments—are telltale. Crystals made in the summer tend to be bigger than winter ones. Ice is a little more acidic when formed from summer snow, dust tends to accumulate in the spring when the winds are stronger, and a dry year means more dust. These pieces of information tell us the year's chronicle of hot and cold, wet and dry. Looking downward at the ice cap samples, we first see nice, even layers, varying in thickness, but easily distinguishable. Farther down, the slipping of glaciers causes small wiggles and then creases. Still deeper, we find layers flattened by the pressure of the ice on top until finally, after 100,000 years of Greenland ice levels, it becomes impossible to distinguish one layer from the next. The record ends. At Vostok, Antarctica, near the South Pole, it stretches further back, as much as 400,000 years, because the dryness of the weather means that only about an inch of snow accumulates each year. The pressure on the lower levels is not as great, but then, of course, the layers are thinner and harder to read.

Ice is such a mediocre heat conductor that colder ice levels stay colder and warmer ones stay warmer. It may seem surprising that ice transmits heat poorly, but the Inuit living in igloos have known that for a long time and our ancestors who stored ice through the summer knew that as well. The deep Greenland ice-level temperature readings retain the memory of their birth for tens of thousands of years, a particularly reliable recall in the case of long hot or cold spells. That chronicle shows a time gone by that was often frigid, with Greenland temperatures falling as much as 35 degrees lower than our present ones.

The data indicate overall agreement with the Milankovitch cycles: warmer temperatures appear roughly 103,000, 82,000, 60,000, 35,000, and 11,000 years ago, but the record also contains surprises, events that may be important in predicting our future. The temperature of the last 10,000 years has been rather even, without big fluctuations, but the ice record shows this is unusual. In the previous 90,000 years, Greenland had more than twenty abrupt temperature changes. They are marked by periods during which snowfall doubled, the amount of dust in the air changed by a factor of ten, and average temperature rose or dropped by almost 20 degrees. Furthermore, these changes occurred quickly, in decades, and sometimes in as little as a few years. The bigger shifts, which lasted for about a thousand years, were preceded by smaller back and forth jumps dubbed climate "flickering." Since the Earth's position relative to the Sun varies very slowly, these global events cannot be explained by Milankovitch cycles.

The abrupt Earth temperature variations are also not due to meteors hitting the Earth or volcanic eruptions sending up clouds of dust, because impacts or eruptions large enough to induce a major climate shift are too infrequent. Something else happened, it happened many times, and it made a very big difference. The last snap was a thousand-year-long cold spell that ended about 11,000 years ago, while the Earth, emerging from an ice age, was entering a warm phase. That transitional millennium is called Younger Dryas after a little flower that spread rapidly as the new European forests became tundra, the Gobi became a desert, and the oceans shifted.

Evidence of the thousand-year Younger Dryas period can be read in Scandinavian glacial moraines, in the movement of New Zealand glaciers, in deep-sea sediments, and in Canadian bogs. We don't know for sure what caused the sudden cold spell and the other temperature snaps that preceded it, but the leading candidate is a shift in the flow of the giant ocean "conveyor belts" that transport water—one of those four main contributors to temperature change—around the globe.

Wallace Broecker, the distinguished Columbia University oceanographer-geologist, has argued this view with increasing vigor. For more than a decade Broecker has been studying the great ocean flow patterns. He's certainly not the first: Benjamin Franklin, for one, was intensely curious about the Gulf

Stream, which his Nantucket cousin Captain Timothy Folger had charted for him. Franklin started by observing birds and fishes; ever the experimentalist, he soon turned to temperature measurements of the air and the ocean. His interest mounted with each successive crossing of the Atlantic. In his final journey, at age seventy-nine, he recorded the ocean temperature at depths down to twenty fathoms, examined the width of the Gulf Stream, and compared it to the surrounding ocean, looking for clues to the comparative warmth of northern Europe.

The great conveyor belts, rather than the Gulf Stream, are the true causes of northern Europe's milder climate. The topography and winds of the Atlantic Basin lead to a northern flow of warm water larger in volume than all the great rivers of North America. On reaching the vicinity of Greenland, the flow is chilled by the Arctic. The colder and hence denser water plunges down and begins a long trip south to the Antarctic. From there it is reflected back into the Atlantic and the Pacific, beginning its journey north once again. Atlantic water is also saltier than Pacific water because of a slightly different balance between evaporation and inflow. The main source of this imbalance seems to be the warm, moisture-laden winds that blow steadily westward across Central America, dropping fresh water on the Pacific and leaving the Atlantic saltier. Salinity affects the conveyor belt because saltier water in the upper layers is denser and hence sinks more rapidly. All this leads us to believe the saltier and more rapidly churning Atlantic is the main driver of the conveyor belt.

Some 13,000 years ago, probably toward the beginning of the Younger Dryas, a dramatic change in the North Atlantic water circulation occurred during a short period—less than 200 years. It seems that the conveyor stopped and then

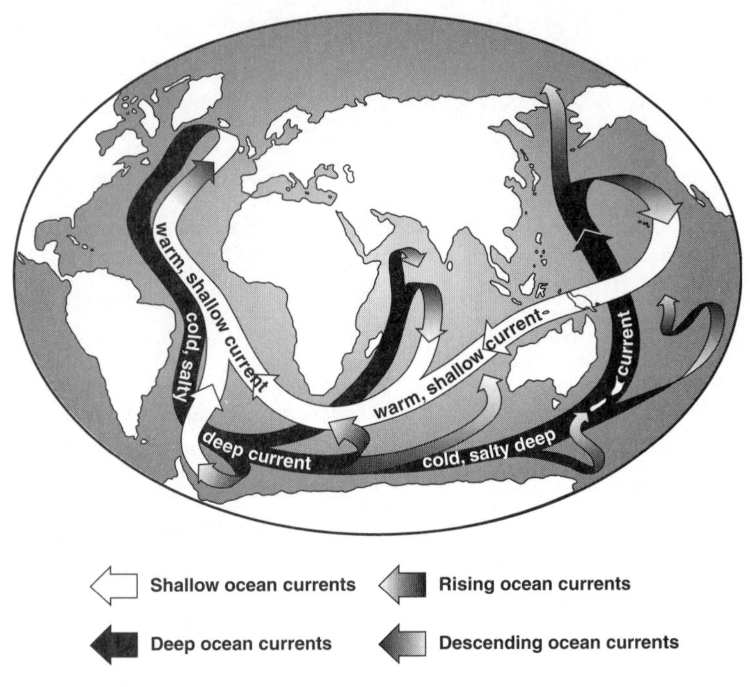

Ocean conveyor belts (after Broecker)

picked up again, at a slower pace. After a thousand-year period, it resumed at full strength. Broecker thinks the cause of the shift can be traced back to the melting of the last ice age's great ice cap; the newly produced water formed an enormous lake in North America, initially spilling over the Mississippi watershed into the Gulf of Mexico. As the ice cap continued retreating, another channel opened to the east along the St. Lawrence River watershed. In Broecker's picture, water began streaming directly into the North Atlantic about 12,000 years ago at a point of entry near where the conveyor belt now reaches its northern end, close to where it sinks to the bottom. He thinks the sudden onslaught of fresh water reduced salinity, decreased the density of the ocean, and stopped the conveyor. The most likely result was a colder Greenland and a warmer Antarctica, since the flow of both warm water going north and cold water going south was halted. Preliminary evidence indicates this occurred. Although more and better experiments need to be done, the conveyor belt model passes the first tests.

Broecker's model is one proposal. Maybe other mechanisms dominated. Ocean sediment analysis suggests that migration southward from the Arctic of great iceberg flotillas occurred at least a half-dozen times in the past 100,000 years. No single system explains all the ocean phenomena nor can one think solely of oceans, neglecting their complex interactions with the Earth's atmosphere and crust.

El Niño, Old and New

Earth-induced, air-induced, and water-induced changes in climate are linked to one another. An event that features ocean-atmosphere interactions has been very much in the public eye recently. Labeled El Niño, the Spanish term for Christ Child, it was given its name by seashore inhabitants of northwestern Peru, who saw warm water currents around Christmastime. About a century ago, these Peruvians also started noticing that periodic changes in water temperature were connected to heavy rains and subsequent floods.

A sudden warm ocean movement in northwestern Peru is an unlikely headline, but its 1997–98 appearance was the El Niño event of the century. The water temperature of the surface of the eastern equatorial Pacific reached a new high every month between June and December 1997, with a final rise of almost 20 degrees. The surface of warm water associated with El Niño was bigger than the continental United States. When the rains came, hundreds were killed in mudslides.

Thirty-foot waves pounded California in February 1998 and Florida had twice the usual rainfall. Crops rotted, and estimates of total lost lives ranged up to 20,000; property damage was estimated at $30 billion. Did El Niño cause all this? Probably not, but the California waves and the Florida rains certainly were El Niño–induced, and those same weather patterns triggered droughts in Asia and forest fires in Indonesia.

At first glance, these different events seem unrelated, but the link is becoming clearer. In 1923 Sir Gilbert Walker first noticed a phenomenon now known as the Southern Oscillation. When the atmospheric pressure goes up in Darwin, Australia, on the Indian Ocean, it goes down in Tahiti, the island in the western Pacific. When it goes down in Darwin, it goes up in Tahiti. The Southern Oscillation is a kind of barometric seesaw between two oceans.

In the late 1960s oceanographers found the Southern Oscillations were strongly correlated to El Niño; the overall phenomenon is now called ENSO, an acronym for El Niño Southern Oscillation. The connection is so tight, the link so clear, that correlations between El Niño storms in Peru and droughts in lands bordering the Indian Ocean have been looked for in the historical record. The most severe drought years in India were 1686 and 1790, with rainfalls dropping to one quarter the usual. To make it worse, rain came only in sporadic torrential downfalls, so rapid that it ran off immediately, leaving the soil still parched. Not surprisingly, the droughts coincided with unusually strong El Niño conditions in Peru.

Also in the 1960s, a Norwegian meteorologist named Jacob Bjerknes noticed that the Southern Oscillation is a perturbation of a general atmospheric east-west flow pattern over the Pacific in which warm and moisture-laden air converges on the western Pacific, typically its warmest and wettest part.

When the air pressure and the trade winds change, circulation does too. Ocean surface temperatures are coupled to airflow patterns, first with a positive feedback mechanism that amplifies El Niño, and then with a negative one that shuts it down. Spurred by the success of these analyses, climate experts are now studying the AO, Arctic Oscillation, the NAO, North Atlantic Oscillation, and other links between ocean movements, air currents, humidity shifts, and temperature changes.

These problems are hard. Climate modeling is advancing rapidly because of bigger and better computers, but the subject is still in its infancy. Computer model predictions of El Niño's onset, peak, and subsequent decay into La Niña—the cold cycle in the Pacific—are still not much more than rough estimates. We know that El Niño appears every three to seven years and lasts for twelve to eighteen months, but we still don't know how big the next El Niño will be. On the other hand, once it starts coming, we have several months to gauge its buildup. The ongoing research has at least given us a chance to prepare and protect.

Archaeological records tell us that the characteristic rainstorms that follow El Niño have battered South America for thousands of years. There are some interesting time variations in that record. Paleoclimatologist Donald Robdell and his colleagues studied sediment at the bottom of Lake Pallcacocha, 13,000 feet up in the Ecuadorian Andes. They drilled a hole twenty feet into the lake bottom, removed core samples, and studied the layers deposited year after year. A wet year means more runoff into the lake from the steep surrounding bowl, while a dry year leaves less dirt on the lake bottom. Initially, they found no surprises, with heavy El Niño rainfalls occurring every three to seven years, just as they do now. However, the older records show some puzzling features. Starting a little more than 5000 years ago, El Niño came only rarely, perhaps a few times a century, and then only in a weakened form.

The Lake Pallcacocha cores go back more than 12,000 years. Clearly more evidence is needed, but there are also hints from corals in Papua New Guinea and Australia that the air and water circulation across the Pacific was different 10,000 years ago. The evidence is tantalizing and perhaps far-reaching for the development of civilization in South America. It's even been suggested that the onset of rainfalls triggered by El Niño led to increased crop production, facilitating the growth of larger specialized communities. In any case, the Andean climate has changed over the course of 10,000 years, perhaps another link to shifts in ocean conveyor belts.

Some climate anomalies are easy to explain and others are not. Each of the two coldest summers in the past 500 years, 1601 and 1816, followed large volcanic upheavals that threw massive amounts of dust into the stratosphere. Over the course of each year, as the slowly spreading dust obscured the Sun, the Earth cooled. As long as an eruption is big enough, the effect works regardless of where on Earth the volcano is located. Tambora, Indonesia, erupted in April 1815, and in Europe, 1816 was known as "the year without a summer." Even the polar ice cap levels of those anomalously cold summers show large quantities of volcanic ash. As recently as 1991, we saw a temporary drop of 1 degree in global temperatures after the explosion of Mount Pinatubo in the Philippines.

The evident correlation between volcanic eruptions and global cooling years is easy to understand. On the other hand, we don't know why the period between 1100 and 1250 was warm enough in Europe and America for the Vikings to grow crops in Greenland. Conversely, 1400 to 1800 was so cool that it is called the "Little Ice Age." Global mean temperatures dropped by a degree or two; the Dutch canals froze over regularly, and the Swedish army invaded Denmark by marching across the iced North Sea. New York harbor occasionally froze, and George Washington experienced a cold winter at Valley Forge—cold but not unusually cold for that era. Were these events due to a shift in winds, a shift in ocean currents, or perhaps even a combination of many factors?

Sunspots also have an effect on our weather, though allegedly too small to induce major climate shifts. However, sunspots seem to have almost disappeared from 1645 until 1715, smack in the middle of the "Little Ice Age." When this was pointed out about a hundred years ago, the results were dismissed as poor record keeping, but recent analyses show the result is right. Furthermore, six out of the seven minima in the Sun's magnetic field over the past 5000 years coincide with intervals of colder climate. It's probably just a coincidence, but there are causes of global temperature shifts we still don't understand that well.

Greenhouse Effect: The Basic Science

Air adds new subtleties to the complex climate-shaping relationship between Sun, water, and Earth, chiefly through the much-publicized and often misunderstood greenhouse effect. The source of heat in a greenhouse is sunlight that has streamed through the glass. All radiation, including, of course, sunlight, can be characterized by its energy density or intensity and its frequency, the latter being the number of waves per second emitted by the sender. Plants soak up the energy and reradiate it at a lower frequency. The air inside the glass house absorbs some of the radiation given off by the plants and heats up. The glass's main function is keeping the enclosed warm air from rising and mixing with its cooler surroundings.

Similarly, and on a much grander scale, our atmosphere acts like a greenhouse. Sunlight comes through the air and, to the extent it is not reflected back by clouds, reaches the Earth and warms it. The Earth reradiates, but the air keeps the heat from escaping. Our own globe's glass cover is the air itself, held in place by the force of gravity. The puzzle is why air acts like a filter, letting radiation from the Sun in but keeping the Earth's radiation from flowing back out into space.

The answer involves a mixture of both physics and chemistry. To understand it, one has to know some of the rules that govern radiation, thermal equilibrium, and temperature as well as the specific properties of molecules in common air. A Viennese physicist named Josef Stefan discovered the first of the relevant radiation rules in 1879. He observed that all objects emit radiation, electromagnetic rays of which visible light is a narrow frequency band, with an intensity from every square foot that depends only on the transmitter's temperature. That dependence is very strong, proportional to the fourth power of temperature. If temperature is doubled, you have 2 × 2 × 2 × 2, or 16, times as much energy density (intensity). Of course, this is only true when temperature is measured with the correct scale; that scale is degrees Kelvin, where zero is –273 degrees Celsius. While all objects radiate, Stefan's rule says the radiation decreases dramatically as the temperature drops, finally vanishing as absolute zero is approached.

Stefan had a very gifted student in Vienna named Ludwig Boltzmann, whom we met in the last chapter. Boltzmann took Maxwell's ideas about the statistical behavior of molecules in thermal equilibrium, recast them, and then introduced the notion of entropy to describe their motion. Stefan led young Boltzmann to Maxwell's papers and lent him an English dictionary to help translate them into German. Boltzmann repaid the favor ten years later by giving a derivation of Stefan's rule based on Maxwell's ideas. The rule is now known as the Stefan-Boltzmann Law of Radiation. It predicts how much radiation is coming out of every square foot or meter of a surface at a given temperature. For example, knowing the surface temperature of the Sun and the extent of the solar surface, we find the total radiation output of the Sun.

However, the Stefan-Boltzmann rule doesn't tell us what frequencies of radiation we should expect. That follows from a rule proposed by Wilhelm Wien in 1894. This says that if you plot radiation's intensity versus its frequency, you will find that the curve has a very sharp peak at a value of the frequency that is proportional to the emitter's temperature, measured in degrees Kelvin. The figure on page 114 shows such a curve for three representative temperatures. Note that the curves can also be displayed as intensity-versus-wavelength plots because frequency and wavelength are inversely proportional to each other. Increasing wavelength corresponds to decreasing frequency.

In discussing how temperature determines the frequency of the radiation intensity peak, the Earth and the Sun form a curious contrast. We don't see the intensity peak of the radiation emitted by the Earth, but we know its location is in the far infrared, because we measure the Earth's temperature. On the other hand, although we don't directly measure the Sun's temperature, we do see the intensity peak of the Sun's radiation. The peak is smack in the middle of the visible range. In fact the human eye surely evolved over millions of years to function optimally in bright sunlight; it's fair to say that the eye was formed in response to the Sun's surface temperature. If the Sun were slightly cooler, our eyes would have adapted accordingly and probably be most sensitive at slightly lower frequencies. The Sun's surface temperature is known to us because of the observed peak in the solar radiation intensity-versus-frequency curve. That temperature is 5800 degrees Kelvin.

The Stefan-Boltzman Law determines the amount of radiation emitted by the Sun once the Sun's surface temperature is known. Then, with the help of a few geometrical considerations, we also know how much of that radiation reaches the

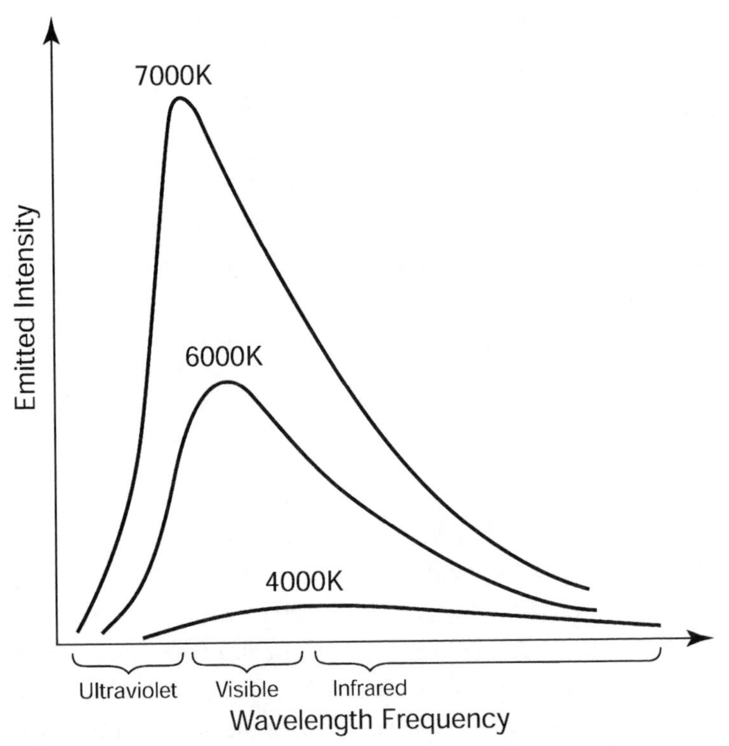

Plot of intensity versus frequency for radiation in thermal equilibrium. Three representative curves are shown corresponding to radiation in thermal equilibrium at three different temperatures: 7000, 6000, and 4000 degrees Kelvin. The shift in intensity and frequency of the peaks as temperature changes illustrates Wien's Law.

Earth. How much does the Earth reradiate? Thermal equilibrium implies the same amounts of heat come in and go out. Knowing how much is reradiated in turn fixes the Earth's temperature. Of course, that temperature has changed a bit from time to time, but those are fine corrections we can ignore right now. Doing the simple calculation of balancing heat in versus heat out shows that the Earth's average temperature is 0 degrees Fahrenheit, or about 255 degrees Kelvin. The correct answer is 60 degrees Fahrenheit. Why the difference?

You would be right if you said the difference was due to the Earth's atmosphere producing a greenhouse effect, but it's not quite that simple. By volume, air is approximately 78 percent nitrogen, 21 percent oxygen, and 1 percent argon. That adds up to 100 percent. All three of those gases are transparent to solar radiation coming in and Earth radiation going out. As usual, the devil is in the details. Those three gases make up almost 100 percent of our atmosphere, but not all of it.

The 60-degree difference between the simple estimate of the Earth's temperature and the real value is almost entirely due to the presence in the atmosphere of small amounts of water vapor, carbon dioxide, and methane as well as traces of other gases. Water vapor is 4 percent per volume of air in the most humid zones of the equator and less than one part per million (ppm) in the Antarctic. Incidentally, contrary to popular image, this excessive dryness means that snow hardly ever falls in the Antarctic. The cold ensures that it doesn't melt when it does fall. This is captured on an old T-shirt belonging to a friend of mine who spent a year in the Antarctic. It says, "Ski the Antarctic—two inches of powder and two miles of base."

While carbon dioxide only makes up 360 parts per million by volume of our atmosphere, it's the focus of a major international debate on global warming. It and other "greenhouse gases" are so important because of their chemistry:

they have a molecular structure that makes them transparent to radiation at the peak of solar emission but highly absorbent of the infrared radiation from the Earth. They are one-way filters, the cause of the 60-degree Fahrenheit rise in the Earth's temperature. They will also be the cause of future increasing global temperatures, if their levels continue to rise.

The detailed argument is a bit less direct because of cloud cover. The thermal equilibrium reached is one between air and Earth—with each of them absorbing and emitting radiation. There's a small amount of heat, small on this scale anyway, produced by radioactivity inside the Earth. But while these factors complicate the argument, the conclusion stands: greenhouse gases are the main cause of the 60-degree Fahrenheit difference in temperature between an Earth with and without an atmosphere.

It may seem surprising that minuscule fractions of our atmosphere should make such an impact on global temperature. It will seem less surprising after looking at Mars and Venus. Although Venus, Earth, and Mars were formed similarly out of a disc circling the Sun, they have evolved on three separate time scales. A series of unmanned vehicles has shown that Mars has a cold, sterile surface. It has little if any water or methane and only a thin, mainly carbon dioxide, atmosphere. That atmosphere is so thin that Mars has a negligible greenhouse effect.

With such a meager atmospheric cover, Mars's temperature swings daily through 200 degrees Fahrenheit. The Martian equator, which reaches a noontime high of 70 degrees, plummets at night to –130 degrees. The Martian poles are so cold that even carbon dioxide freezes. The traditional view says Mars once had abundant water and perhaps even life. Volcanism ended, the planet cooled, and the water that did not escape ended in deep, buried frozen reservoirs. That's the traditional view. Very recent pictures from the Mars Global Surveyor hint at volcanism and possibly even water flowing on the surface not that long ago, so the traditional view may change in the coming years.

In contrast to Mars, robotic missions exploring Venus reveal an overheated planet. If Earth and Venus resembled Mars more closely, their average surface temperatures would all be near 0 degrees Fahrenheit. The greenhouse effect raises Mars's temperature by a few degrees, Earth's by a comfortable 60, and Venus's surface by 800 degrees, making it hot enough for rocks to glow and metals to melt. Venus has volcanism and radioactivity in its crust, but only at levels comparable to those on Earth. The source of the extraordinary temperature is something else, a runaway greenhouse effect. Venus has too much carbon dioxide and too little water vapor. From a geocentric point of view, we say Mars has too little greenhouse effect and Venus has too much.

Abundant water and perhaps even life were once present on Venus, but they were destroyed. Our best guess is that water vapor drifted upward as the once much cooler planet heated up. In rising, the vapor encountered strong solar ultraviolet radiation that disassociated the water molecules into hydrogen and oxygen. The light and fast hydrogen atoms drifted off into empty space over the course of 600 million years until the water clouds were gone. A new thermal equilibrium was eventually reached, one in which the surface temperature of Venus attained its current 800 degrees.

Planets have a certain kind of stability, at least for a while. Imagine an enhanced Earth greenhouse effect. The rise in temperature increases evaporation from the oceans, producing a greater cloud cover. This lowers the temperature—a self-correcting feedback mechanism is at work. Living organisms provide another feedback mechanism. Higher levels of carbon dioxide in the air means more of the gas dissolves in ocean water; this forms an acid that combines with calcium to make calcium carbonate, the stuff of seashells. Their increased production helps keep in check the levels of carbon dioxide. As long as feedback mechanisms keep the balance, the Earth won't follow Venus's path.

A billion years from now, solar dynamics ensure that our Sun will brighten enough to launch us on a runaway mode, ending water vapor and life. This will leave the Earth looking a lot like Venus does today. The journey of the three planets started a little more than four billion years ago, and ended in fire for Venus and ice for Mars. As Robert Frost wrote in his poem entitled "Fire and Ice":

> Some say the world will end in fire
> Some say in ice
> From what I've tasted of desire
> I hold with those who favor fire.
> But if I had to perish twice,
> I think I know enough of hate
> To say that for destruction ice
> Is also great
> And would suffice.

The Earth will end in fire, certainly in a billion years. Of course, that assumes we haven't found in the meantime a way to move the Earth a little farther out in the solar system, but more on that later.

Greenhouse Effect: The History

Not much was known about air until 1750. Before then, "subtle matter" or "ether" was presumed to fill the universe, but the new emphasis on measurement and the development of the tools of analytical chemistry made scientists ask again about air. Carbon dioxide was identified by Joseph Black in the 1750s, nitrogen by Daniel Rutherford in 1772, and oxygen by Carl Scheele and Joseph Priestley a few years later. In 1781 Henry Cavendish determined that air was 79 percent nitrogen and 21 percent oxygen, values close to our modern ones. He missed finding the 1 percent of argon. That took another hundred years because argon, a noble gas, is chemically inert.

Cavendish was a great eccentric. The nephew of the Duke of Devonshire, he eventually became, through inheritance, one of the richest men in Britain. Cavendish led a solitary life, with no social life except for an occasional visit to the Royal Society. His clothes were shabby, he never married, and is described as having "an air of timidity and reserve that was almost ludicrous. . . . His dinner was ordered by a note left for his servants on the hall table and his women servants were ordered to keep out of his sight on pain of dismissal." Cavendish never finished his degree at Cambridge, his papers were mainly unpublished, but his brilliance gradually came to be appreciated. His descriptions sound quaint: "common air" (air), "inflammable air" (hydrogen), "fixed air" (carbon dioxide), and "phlogisticated air" (oxygen), but the differentiations are valid and the experiments that back them up are ingenious. Cavendish even realized that combustion required "phlogisticated air," while too much "fixed air," as produced by fermentation or rotting, would snuff the flame out. This may be the first recorded measurement of high carbon dioxide levels.

The greenhouse effect is first mentioned in a paper by the great French mathematician Jean-Baptiste Joseph Fourier. He worked for many years on the problems of thermal conduction, summarizing this research in his 1822 *Analytical Theory of Heat,* a book that remained a classic for well over a century. A few years after its publication, Fourier turned his attention to the role atmosphere plays in trapping heat emitted by the Earth. In comparing the effect to the air inside a glass vessel, he was the first to employ the analogy to a greenhouse.

Fourier's analogy was useful. However, he didn't study which gas in "common air" traps the heat. The physics of thermal conduction and radiation were first combined with the chemistry of air by John Tyndall, well known for having added rigor to Agassiz's studies of glacier flow. In the late 1850s Tyndall turned his attention to a systematic examination of the properties of air. He found that nitrogen and oxygen are essentially transparent to both incoming solar and outgoing Earth radiation while water vapor, carbon dioxide, and methane absorb reradiated infrared rays. Tyndall's familiarity with Agassiz's notions about past glacial ages then set him to thinking about what might have caused those ice ages. He speculated that shifts in the levels of carbon dioxide, water vapor, and methane might be responsible, as he stated, for "all of the mutations of climate which the researches of geologists reveal."

Cavendish to Fourier to Tyndall. The next big step was taken in the 1890s by the Swedish physical chemist Svante Arrhenius. While Tyndall thought carbon dioxide, water vapor, and methane could lead to global warming, Arrhenius wanted to determine how much global warming they would cause. His motivation, at least in part, was the same as Tyndall's—seeing if past ice ages might be due to lower levels of greenhouses gases. Arrhenius attempted to calculate the expected rise in global temperature from growing levels of carbon dioxide. Taking into account the feedback effects of water vapor, Arrhenius predicted that doubling the amount of carbon dioxide in our atmosphere would lead to an average global temperature increase of 10 degrees Fahrenheit. This result, now about a hundred years old, is the first true estimate of global warming due to greenhouse gas emission. At the time, it wasn't considered a cause for any alarm, particularly in Arrhenius's native Sweden. If anything, a warming trend was thought of as a good thing, likely to lead to more favorable crop growing and living conditions.

That's largely where matters stood until the 1950s. Most scientists thought Arrhenius had vastly overestimated global warming. He hadn't realized that atmospheric carbon dioxide was likely to be absorbed into the surface levels of ocean waters and then rapidly mixed through all levels of ocean water. However, as Roger Revelle, then director of the Scripps Institute of Oceanography in La Jolla, California, began to study the question, he saw that mixing was overestimated. According to Revelle's observations, 80 percent of the carbon dioxide added to the atmosphere was likely to stay there. Revelle described this increase as "Man's first great global geophysical experiment" and sounded the alarm.

By the late 1950s, Charles Keeling, one of a small group of scientists concerned about global warming, started a continuous monitoring of atmospheric carbon dioxide levels. Gilbert Plass, another one of the group, published an article in the July 1959 issue of the *Scientific American* predicting that world temperature would rise 3 degrees by the end of the century. The magazine captioned the article with a heading, "Man upsets the balance of natural processes by adding billions of tons of carbon dioxide to the atmosphere each year."

Today, global warming has been elevated to the status of an international crisis. One of the charges to the United Nations–established Intergovernmental Panel on Climate Change is to review the scientific evidence and assess the dangers of global warming, reporting its results to the world public every five years. Their latest document, a thousand pages long, is the third one issued since 1990. Written by a panel of forty-five experts, it is reviewed by hundreds more and presented to the 150 nations participating in the document's preparation.

Looking at population growth, economic development, and technological evolution, the 1997 Panel came up with three estimates of the rise in atmospheric levels of carbon dioxide over the next hundred years. The present 360 ppm (parts per million) level was compared to 277 ppm, the approximate level in 1750. That date was picked because it marks the beginning of both the Industrial Revolution and the first large-scale deforestations, each of which set carbon dioxide levels on their quickening slope. The optimistic Panel projection for 2100 was 450 ppm, a mid-range scenario led to 700 ppm, and the worst-case estimate was 954 ppm. On this basis, the Panel thought the average world temperature would go up by 2 to 6 degrees Fahrenheit over the course of the next hundred years and ocean levels would rise anywhere from six inches to three feet.

The latest Panel report, prepared for the end of the year 2000, revised upward the worst-case scenario, now predicting a temperature rise of between 6.3 and 11 degrees from 1990 levels.

Everybody agrees worldwide carbon dioxide levels are going up. The question is how far will they go? The rise depends on how fast the world population grows and, more importantly, on how that population lives. Higher temperatures increase evaporation from the oceans, which means more heat-absorbing water vapor in the atmosphere and also greater cloud cover. Clouds reflect sunlight, cooling the Earth. Carbon dioxide–induced global warming melts polar snow-

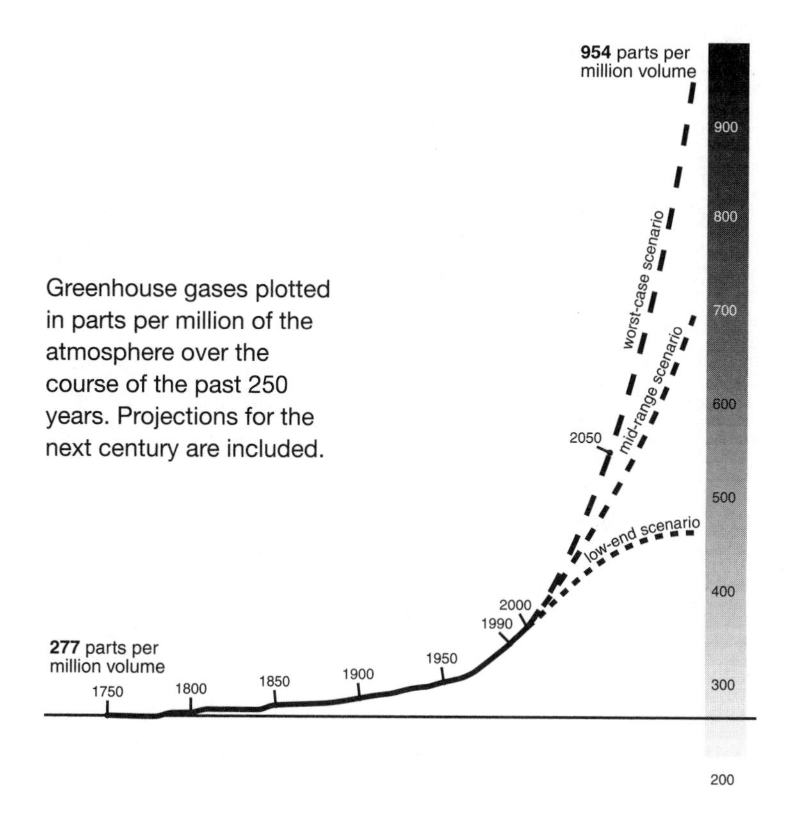

Greenhouse gases plotted in parts per million of the atmosphere over the course of the past 250 years. Projections for the next century are included.

fields, but forests that absorb carbon dioxide might replace those snowfields. On the other hand, forests reflect less sunlight than snowfields. To further complicate the matter, very recent estimates indicate that global warming reduces the beneficial carbon uptake by vegetation. We do know that carbon in the atmosphere leads to carbon dioxide. Unfortunately, we don't know the location of all the carbon. Approximately a million million kilograms of carbon, some 20 percent of the total carbon that is reabsorbed each year, is missing. We don't know how it left or where it went. It might be disappearing into giant Siberian forests or some not well understood oceanic processes. These are all pieces in the puzzle, parts of the picture we can see only dimly. We don't know how much atmospheric carbon dioxide there will be ten years from now, much less a hundred years from now, but past experience does not lead us to be optimistic.

There are hundreds of studies on the subject, most of them needed and important. Each study examines a piece of the picture and presents its results with associated uncertainties. A meaningful estimate of our future requires interconnecting all the pieces, forging the links, and then extrapolating. Although many unknowns make planning hard, we still need to seek the answers. Doing nothing is not a viable alternative. Some models estimate that the worst-case scenario for 2100 is an average global warming of 14 degrees, 3 degrees higher than the Intergovernmental Panel on Climate Change's worst case. Nobody says this isn't a serious problem for us all.

What should we do? Carbon dioxide is a natural byproduct in the burning of all fossil fuels, such as coal and oil. In the distant future, the combined use of nuclear energy and new renewable energy sources may allow us to be much less dependent on coal and oil. But in the near future this won't happen. Carbon dioxide levels in the atmosphere will continue to grow and continued global warming seems inevitable. Limiting the use of fossil fuels is clearly difficult and presents severe political problems. So again, what should we do?

Dr. James Hansen, director of the National Aeronautics and Space Administration's (NASA's) Center on Space Studies, has a suggestion. He has been recognized as one of the world's experts on this issue ever since he published his influential 1981 paper on global warming's connection to rising levels of carbon dioxide. When he recently expressed some optimism on the subject, his remarks received a great deal of attention. Hansen says the problem of stringently limiting carbon dioxide emissions is too hard for the time being. We should of course make efforts to curb emissions, but too much emphasis has been placed on carbon dioxide without enough attention to the other greenhouse gases. Hansen claims most of these are easier to control and can have their production cut without significantly affecting the quality of life. Each one of them is relatively unimportant, but their total contribution is not.

Methane is a good example. Now present in the atmosphere at levels of about 1700 parts per billion, it's produced by a variety of unpleasant means: leaking pipelines, poorly drained rice paddies, and belching, farting livestock. A partial remedy would involve feeding antigas additives to New Zealand sheep, having Chinese rice paddies drained more efficiently, and making sure the Alaska pipeline is serviced regularly. It's not a complete solution, but every little bit helps.

Nitrous oxide is also important, as are the chlorofluorocarbons produced by aerosol cans or refrigerants. Rare gases like sulfur hexafluoride or the recently discovered trifluoromethyl sulfur pentafluoride are only present in the atmosphere in parts per trillion. Although this might seem insignificant, one part of sulfur hexafluoride contributes as much to global warming as 24,000 parts of carbon dioxide. These gases are rare, even extremely rare, but they act powerfully to warm the Earth.

Hansen goes on to emphasize the importance of other contributors to global warming. Soot, a dust and not a gas, is a significant factor in smog. After 4000 people died in the killer London smog of December 5–9, 1952, English legislation was passed to regulate it, but soot remains a major problem in developing countries. Ozone, the molecule formed from three atoms of oxygen, is a crucial presence in the stratosphere, stretching roughly from six to thirty miles above the Earth. It absorbs the harmful ultraviolet radiation from the Sun. On the other hand, ozone near the Earth is a worldwide problem, contributing to global warming.

Hansen feels a systematic attack on everything but carbon dioxide will buy us some time while we work on developing alternative sources of energy. John Holdren, professor of environmental science and public policy at Harvard, thinks that strategy is a mistake. "This isn't an either/or problem. It's a both/and problem. We are going to need all the cuts we can get."

Global warming may not even be the biggest of our climate worries. Broecker, the pioneer in studying ocean conveyor belts, chose "Chaotic Climate" as a title for a *Scientific American* article of his because he thinks the prevailing weather might change unpredictably once a threshold is passed. He speculates we might run into a Younger Dryas shift with temperatures in northern Europe dropping precipitously, Scandinavian forests becoming tundra, and the green grass of Ireland disappearing. To make matters even worse, Broecker thinks that the flickering that would precede such a shift might be more disruptive than the shift itself.

Greenhouse Effect: The Politics

The importance of carbon dioxide in producing a greenhouse effect was highlighted at the Earth Summit held in Rio de Janeiro in 1992. At that meeting, the richer industrialized nations adopted a voluntary goal of limiting their greenhouse gas emissions to 1990 levels by the year 2000. The 1997 Kyoto Conference tried to reaffirm that principle and specify further reductions to be reached by the years 2008–2012. The basic aim in setting these limits was to stabilize the carbon dioxide levels in the atmosphere at roughly twice what they were before the Industrial Revolution.

The Kyoto Protocol has not been ratified by any of the world's major industrial nations. It stipulates that, by 2012, there should be a 5 percent reduction in carbon emissions below their 1990 levels. The follow-up meeting to the Kyoto Conference took place in November 2000 in The Hague. With agreement in sight on carbon emission levels, this accord foundered on the extent to which forests could be counted as carbon sinks, balancing in part a nation's carbon emissions. The United States' opening position was that its forests should be calculated as contributing 300 million tons of carbon absorption per year. That request was reduced at the last minute to seventy-five million tons per year, but the only number acceptable to Green Party European ministers was zero. That's where the meeting ended, with bitterness on all sides. Dr. Michael Grubb, a professor of climate change and energy policy at London's Imperial College, was quoted afterward as saying, "When something like this is killed, it is killed by an alliance of those who want too much with those who don't want anything."

The United States' level of energy consumption, the highest in the world, angers many Europeans. They see Americans driving larger and larger vehicles, with gasoline prices kept artificially low. They resent President George W. Bush's March 2001 announcement that the United States would withdraw support for the Kyoto Protocol, saying its binding limits on emissions could hurt the U.S. economy. Some European governments have found unplanned ways to approach the Kyoto Protocol limits. The union of the two Germanys led to the closing of the old heavily polluting industries in East Germany and a reduction of greenhouse gas emissions. The union-breaking tactics of Margaret Thatcher's government forced many British coal mines to shut down and made the British switch to natural gas, the so-called dash for gas. This too reduced the production of greenhouse gases.

The United States is becoming increasingly isolated in its opposition to the Kyoto Protocol. A 2001 U.S. governmental suggestion that more research was needed to determine the causes of global warming, with an inherent challenge to the Intergovernmental Panel on Climate Change, has been refuted by a White House–appointed panel of the National Academy of Sciences. Their June 2001 report, while agreeing that more research was desirable and even necessary, agreed with the IPCC's projections of greenhouse-gas-induced temperature rises. After an international agreement was reached in July 2001 to uphold the Kyoto Protocol, the *New York Times* ran a story with the headline, "178 Nations Reach a Climate Accord; U.S. Only Looks On."

Beyond the struggle of developed nations with one another, there is also a rising split between them as a group and the developing nations, a battle between the haves and the have-nots. The United States Senate backed unanimously a resolution in 1997 introduced by Democratic senator Robert Byrd from West Virginia advising the administration not to sign any treaty unless poorer nations agreed to limit emissions on the same timetable as the richer countries. The world's weaker countries see this kind of action as a new colonialism, an effort by the world's economically stronger countries to prevent them from developing industrially and technologically—a top-down preservation of the balance of economic power. Specifically, the G-77—a group of seventy-seven developing countries including India, the African states, most of Latin America, and much of the Middle East and Southeast Asia—opposes reduction of greenhouse gas emissions on the same timetable as the industrialized countries. They feel the rich countries created the problem and that they need to take the lead in solving it. But tensions abound even in the G-77: oil-rich Venezuela, Kuwait, and Saudi Arabia would like to see continuing high levels of oil consumption. Bangladesh, so vulnerable to rising ocean waters, fears oil-consumption-induced global warming will lead to massive flooding of its low-lying shores. There are impasses within impasses in reaching an agreement.

In September 2000, an article entitled "Equity and Greenhouse Gas Responsibility" appeared in *Science* magazine. The title caught my eye and, frankly, so did the fact that one of the eight authors was John Harte, an old friend of mine. We were physics students together and met again as postdoctoral candidates, but his interests shifted as it became clear to him that scientific research on environmental issues was going to become increasingly important. The claim made in the article by John and his coauthors is important: that all people have equal claims to common resources. The atmosphere and the oceans do not belong to one country, and we should share equally in their wealth and contribute equally to their maintenance. After hearing this argument, one United States official was quoted as saying, "To me this is global Communism. I thought we'd won the Cold War." Some clearly view this philosophy of equal claims as questionable, but the United Nations Convention on the Law of the Sea dictates common ownership of all deep-sea resources and, by implication, responsibility for the health of the oceans.

There are many problems in reaching a goal of equal sharing of resources and equal responsibility for their maintenance. It seems politically unfeasible. At present, the United States per capita carbon emission average is more than five metric tons per year, while over fifty developing countries have emissions under 0.2 metric tons, 4 percent of the U.S. average. Having greenhouse gas emissions kept to less than twice preindustrial levels for a projected world population of ten billion means limiting per capita carbon emissions to a global average of 0.3 metric tons per year. This target seems totally unrealistic for developed countries. Nevertheless, in the long run, the principle of equity is the only ethically justifiable principle, and the political framework for setting it in place has to include environmental matters and, in particular, the uses of air, earth, and water.

The Sun will continue to treat us all equally, as it has always done.

Chapter 4

LIFE IN THE EXTREMES

STARTING IN THE 1750s—it's not known how or why—the spirit of the sea came to be known in the British Navy as Davy Jones. If a sailor died on board, he was wrapped in a cloth, taken to the ship's edge, and then, after a brief prayer, heaved overboard to rest forever on the cold, dark, ocean bottom, in Davy Jones's locker. If the sailor had drowned and his corpse was recovered, it was simply sent back to the water. As Robert Lowell says in "The Quaker Graveyard in Nantucket,"

> The corpse was bloodless, a blotch of reds and whites,
> Its open, staring eyes
> Were lustreless dead-lights
> Or cabin windows on a stranded hulk
> Heavy with sand. We weight the body, close
> Its eyes and heave it seaward whence it came.

The ocean floor on which the body comes to rest is usually monotonously flat with gentle rolling hills coming one after another in a regular succession. One underwater explorer, Cindy Lee Van Dover, has described it as looking much like a Midwestern prairie. It's cold at the ocean bottom everywhere, less than 40 degrees Fahrenheit at depths greater than 3000 feet. It's coldest near the Poles, where salt barely keeps the water from freezing. The average ocean depth is 12,000 feet.

The pressure of the overlying water is enormous—some 300 times as great as that exerted by air at the Earth's surface. Pressure is doubled just going thirty feet down, explaining why it is so difficult for humans to dive to depths of more than a few hundred feet.

It's also totally dark on the deep bottom. Sunlight filters through the upper layers of the ocean, but by a thousand feet down there isn't enough light to sustain photosynthesis, the process by which plants convert water and carbon dioxide into oxygen and organic material. Plant growth ceases at that point, but life continues, sustained by nutrients that have drifted slowly downward from the upper, life-rich layers. Food is cycled and recycled by organisms constituting a chain joining the top of the sea to the bottom. Some life, even varied forms specially adapted to darkness and pressure, is sustained by the relatively slim pickings of that muddy ocean floor. It hardly seems like a propitious place for great discoveries. Yet some of the most interesting insights into the origins and the variety of life on Earth have come from those ocean floors; there, high temperatures on floor vents form unexpected environments for the re-creation of the creation. The discoveries on the vents, made only recently because of the difficulties in reaching great depths, have shown us a bewildering array of strange creatures that live in total darkness at high temperatures, temperatures generated by the drift of continents and the great sources of energy buried deep in the Earth. The story is still far from complete. Its uncovering began through the efforts of two enterprising explorers.

Barton and Beebe's Bathysphere

By the time of World War I, the great voyages of exploration of the Earth's surface were ending. Both Poles had been reached, the jungles traversed, and even the conquest of Himalayan peaks was well under way. Two great new frontiers emerged: space in and beyond the Earth's atmosphere, and the deep sea. The former had been explored for some time with balloons and, after the Wright brothers' flight, by mechanized devices. The ocean bottom, challenging and fascinating, remained unexplored but not ignored. At the end of the nineteenth century, Jules Verne's *Twenty Thousand Leagues Under the Sea,* with the *Nautilus* submarine piloted by the legendary Captain Nemo, was being read by youngsters everywhere, and interest in this new frontier was aroused. But, below 1500 feet, submarines risk having their shells collapse from the external pressure. Pressurized suits and metal helmets help divers, but humans can't descend more than a few hundred feet below the surface in such suits without the pressurized air becoming toxic.

In the late 1920s, two New Yorkers named William Beebe and Otis Barton began building a device for deep-sea exploration. They called it a bathysphere, from the Greek *bathy* for "deep"; simply put, it was a sphere intended to go deep. A little less than five feet across, the bathysphere was made of inch-and-a-half-thick steel with two windows of three-inch-thick fused quartz fitted into special viewing ports on either side. There was barely room inside for Beebe and Barton, two canisters of oxygen, a tray of soda lime to absorb the carbon dioxide they exhaled, and a tray of calcium chloride to soak up moisture. Weighing 4500 pounds, the sphere was lowered and raised by a seven-eighths-inch-thick steel cable connected to a steam winch. This was operated on board the accompanying barge, the *Ready.* In order to have a light source to illuminate the dark ocean floor, Barton and Beebe ran a hose containing an electric wire and a telephone wire along the cable joining the sphere to the barge. This relatively primitive arrangement

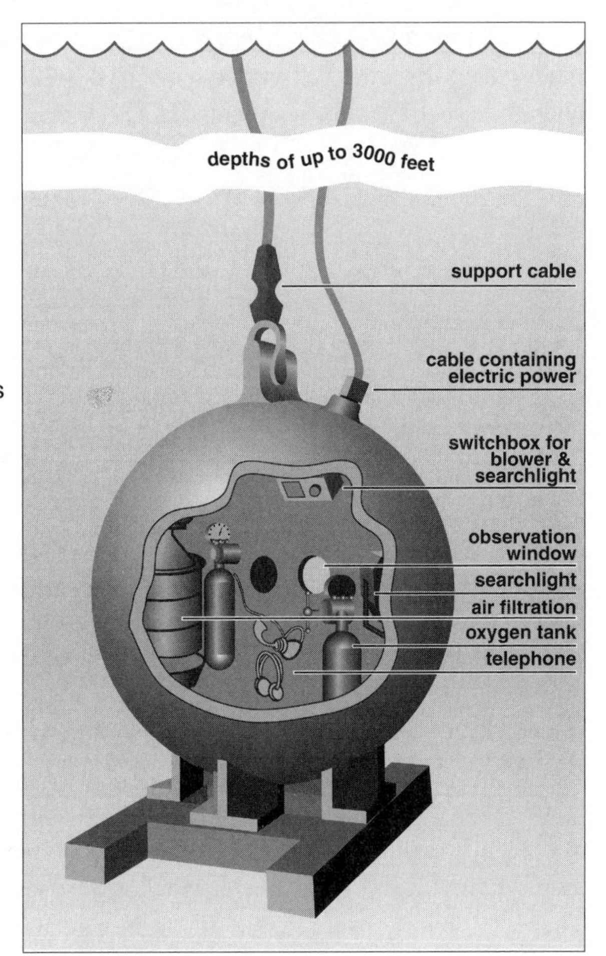

Schematic picture of the bathysphere's interior

allowed the wire to provide current that powered a 250-watt spotlight aimed out the portholes. The telephone allowed Beebe and Barton to communicate with the surface.

In June 1930, the bathysphere dove 1400 feet down. By 1934, Beebe and Barton reached their record depth of 3028 feet, as far down as their cable and winch would allow. In the dim spotlight they saw many strange and wonderful creatures.

They occasionally got carried away, reporting on a mixture of truth and fantasy—what they had seen and what they thought they had seen. Beebe, somewhat of a showman, once used the telephone to make a live radio broadcast on NBC from the ocean bottom. His 1934 book *Half Mile Down* describes "abyssal rainbow gars, sabre-toothed viperfishes, . . . gleaming-tailed serpent dragonfishes and exploding flammenwerfer shrimps." These sensational reports stimulated the quest to know more about what lay on the ocean floor.

The next major pioneer of the depths was a daredevil Swiss physicist named Auguste Piccard. Motivated by a desire to study cosmic rays—particular forms of radiation that reach the Earth from space, attenuated in intensity as they descend through the atmosphere, requiring going to high altitudes to see them—Piccard made his first exploits upward journeys, in a hydrogen-filled balloon. He reached a record height of 55,800 feet in 1932. Paradoxically, the ascents triggered his thinking about descents to the ocean bottom. Realizing the limitations of Beebe and Barton's vehicle, Piccard conjectured that a balloonlike device, which releases a light fluid to sink and jettisons a ballast to rise, might work—no winches and no cables. Piccard started planning a new device, a downward-traveling balloon. He called it a bathyscaph.

With the intervention of World War II, fifteen years passed before Piccard's bathyscaph was ready to dive. A scaled-up version of the bathysphere, it was a seven-foot-diameter steel sphere with three-inch-thick walls and six-inch-thick Plexiglas windows. Still only carrying two passengers, the sphere's novelty was its attachment to a fifty-foot-long flotation tank filled with gasoline to provide buoyancy as it descended. In effect, the flotation tank acted like a hydrogen balloon, raising the observation capsule from the ocean bottom instead of from the ground. To ascend, the bathyscaph simply released its ballast, a series of small iron spheres. In late September 1953, the by then seventy-year-old Piccard and his son Jacques descended in the *Trieste I* bathyscaph to the bottom of the Mediterranean, reaching a depth of over 10,000 feet. Seven years later, Jacques Piccard and Lieutenant Don Walsh of the United States Navy descended, in the *Trieste II,* to the bottom of the Marianas Trench in the Pacific near Guam. They reached the deepest bottom, 35,800 feet down—the oceanic counterpoint of Mount Everest.

Much of the raison d'être for these epic dives was the spirit of adventure, but by the 1960s, a new reason for wanting to study the ocean floor was beginning to emerge. Some fifty years earlier, Alfred Wegener, struck by the congruity of the eastern shore of South America to the western shore of Africa, proposed that they were once joined. He then generalized the idea into a scheme now commonly known as continental drift. Bringing the continents together in the proper fashion using maps, he found that the mountain chains on one continent matched those on another. As he said, "It is just as if we were to refit the torn pieces of a newspaper by matching their edges, then check whether or not the lines of print run smoothly across. If they do, there is nothing left but to conclude that the pieces were in fact joined this way." In this manner, by cutting and pasting, Wegener came to the conclusion that all the continents had once been joined into Pangaea, the name he gave to the single landmass.

Lacking a plausible mechanism for the motion of continents, the theory was at first dismissed. Now it's accepted. Ultimately, the energy source that drives all the motion is now known to be the heat that almost liquefies the deep rock layers at the base of the Earth's crust. By fifty miles down, the temperature is 2000 degrees Fahrenheit; there is enough heat to create volcanoes, enough to sustain the motion of continents and of ocean floors. We can see the effect of heat and high temperatures on rocks in volcanoes, but despite their dramatic appearance, this manifestation of thermal activity is trifling compared to its movement of the entire Earth's crust.

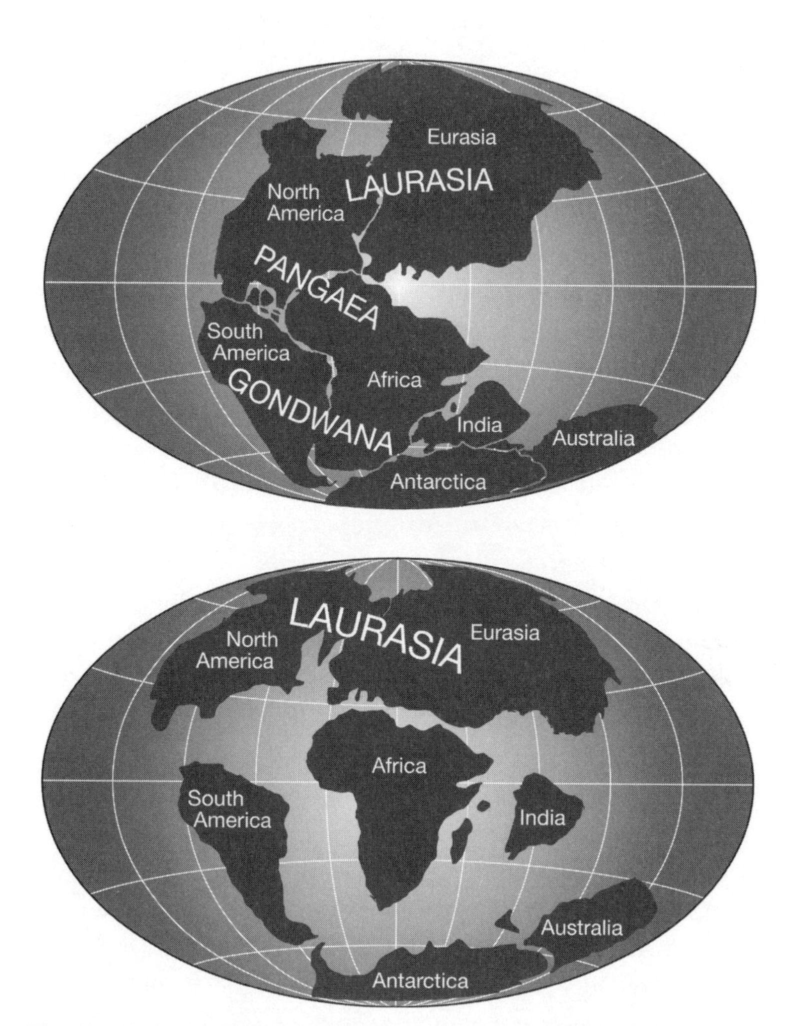

The shaping of the continents due to plate tectonic movement. This illustration shows how the plates moved over the period ranging from 200 million years ago (top) to 100 million years ago.

Our planet's outer shell, varying in thickness between forty and seventy miles, is made up of seven giant slabs and about twenty smaller ones squeezed between them. These slabs or plates rest on thicker strata of semimolten rock that move in giant loops, rising, moving parallel to the surface for a while, cooling, and then sinking, only to resume their motion again. As semimolten rock glides along, it carries with it the overlying plate. The middle regions of the plates are fairly tranquil, geologically speaking, but the boundaries, where two plates either collide or separate, open up a window to the activity of the world below.

Mountain ranges form as plates squeeze together or as one slides under the other, lifting it up on its back. They can also be formed as plates move apart; when they do, semimolten rock from deep within the Earth streams up to fill the rift, forming a ridge as it cools and solidifies. Rifts can develop on solid land, rupturing the crust on top as they emerge; rift valleys in East Africa are an example.

The most spectacular rifts are in the spreading of the ocean floor. They have a very characteristic structure, displaying itself as a long mountain range that may be a thousand or more miles across and a mile or more high. The range, with a central valley, or rift, spreads symmetrically away from the central rift at an inch or more a year. There are two particularly pronounced examples of such structures, one in the Pacific and one in the Atlantic. The former originates near California, goes down the Pacific near the Galápagos Islands, continues southward, and then bends west, continuing between Australia and Antarctica until it joins up with the Indian Ocean ridge. The Mid-Atlantic ridge goes mainly south, stretching from Greenland to below the tip of South America.

These ridges are typically between 5000 and 10,000 feet underneath the surface, too deep for submarines to study and unsuitable for exploration by unmaneuverable bathyscaphs. In the 1960s, new technology allowed a navigable underwater vehicle, known as a submersible, to be built. Just as the bathyscaph resembles the bathysphere, the submersible resembles the bathyscaph, with two key improvements. First, there's no flotation tank. Buoyancy comes from filling much of the space in the hull with syntactic foam, a substance that is both light and resistant to pressure. The second great improvement is the maneuverability of the submersible, made possible because it isn't tied to a large flotation tank. The submersible has a rear and two side propellers, allowing it to move on the ocean bottom, to look and explore. It also has something else that bathyscaphs don't have, a remote-controlled pincer arm to grasp objects. A submersible can see something, move around to scoop it up, and then bring it to the surface.

The first submersible was christened at Woods Hole, Massachusetts, in June 1964. It got its name, *Alvin,* as a contraction of Allyn Vine, the respected geophysicist who had pioneered its development. Very soon after its christening, *Alvin* proved its use to the U.S. Navy, which had sponsored its development. In a midair refueling accident off the coast of Spain in 1966, a B-52 bomber broke in half. The crew of the bomber parachuted out as the wreckage fell. (The crew of the refueling plane was less fortunate.) The wreckage from the B-52 bomber contained four, fortunately unarmed, hydrogen bombs. Three fell on land near the coast and were quickly recovered, but the fourth landed in 2500 feet of water. As panic spread in southern Spain, *Alvin* was shipped to the site. On its twentieth dive, it found the bomb. More than thirty years later, *Alvin,* the workhorse of underwater exploration, still makes 150 dives a year. As it does, it continues to earn the scientific community's gratitude and reveals new and exciting features of the ocean bottom.

In the early 1970s, the two leaders in deep-sea exploration were France and the United States. At that time, they began collaborating on Project FAMOUS, or French-American-Mid-Ocean-Undersea-Study. The French had the *Cyana* submersible and the United States had *Alvin*. The purpose of the collaboration was to study ridges and rifts, the underwater mountain ranges and seams on the Earth where plates were moving apart. In 1974 the teams were ready to study the Mid-Atlantic rift, located at an average depth of close to 10,000 feet. *Alvin* had originally been designed to go no deeper than 6000 feet, but a new titanium covering allowed it to dive twice as deep, more than sufficient to navigate comfortably around the rift.

The Mid-Atlantic plates, separating at about an inch a year, open the ocean floor to the heat from below. However, in the first mid-1970s explorations, the divers were puzzled because the temperatures along the rift were a little lower than they had expected. They thought they might see some real geysers shooting up, but there didn't seem to be any.

Clambake I: Hydrothermal Vents

In 1975, plans were set in motion for *Alvin* to explore a rift about 400 miles west of Ecuador, near the Galápagos Islands. The spot, an eastern spur of the Pacific ridge, is known to be at the location of two separating plates, Cocos and Nazca. Compared to their counterparts on the Mid-Atlantic rift, the plates are moving apart very fast, a promising sign in the search for ocean bottom hot springs. Temperature anomalies in the water had been measured there in 1976. An underwater camera revealed mounds of clamshells that looked as if they had been thrown from a boat after a big dinner, and the proposed dive area came to be known as Clambake I.

Alvin in the deep ocean

In February 1977, *Alvin* made its first dive over the rift. At first the terrain seemed similar to that encountered in the Mid-Atlantic, a barren field of lava. As *Alvin* approached Clambake, the water warmed to upward of 60 degrees. Suddenly it was teeming with life. Bob Ballard, the former director of the Woods Hole Center for Marine Exploration, was on board *Lulu*, the support ship, when the pilot, Jack Donnelly, took two crew members, Jack Corliss and Jerry Van Andel, down on that dive. Ballard describes the scene:

> Corliss was looking at clams—giant specimens, measuring a foot or more in length. And that was only the beginning. He and Van Andel stared in amazement as shrimps, crabs, fish, and small lobster-like creatures passed their viewports. Corliss recognized a pale anemone. He could not identify the weird stuff growing on the bottom, like dandelions, or the wormy stuff attached to rocks.

The "wormy stuff" was clusters of large tubeworms, up to ten feet high. The "dandelions" turned out to be a new kind of jellyfish. The expedition, not expecting to find life on the rift, had not included any biologists on the trip. Nor had they brought along fluids for preserving specimens. Improvising, they used whatever alcohol they could find on board. The "weird stuff" was brought back to Woods Hole in vodka by a sober, but excited, crew.

The dive along the Galápagos rift confirmed the existence of hot springs seeping up through cracks in the ocean floor. The highest temperature recorded by *Alvin* in the early dives on the Galápagos rift was 73 degrees Fahrenheit. Mineral deposits on the ocean floor suggested there had to be plumes of much hotter water since minerals would not have dissolved underground at 73 degrees.

Much hotter water was found in 1979 when a group from Woods Hole took *Alvin* to explore a rift found 9000 feet underwater near the coast of Mexico. This time biologists came along. The ocean floor was spreading particularly quickly along that ridge, a good place to look for vents like the ones spotted near the Galápagos and perhaps even hotter ones. In late April of that year the first dive took place. *Alvin* was maneuvering along the bottom at the new location when it ran into a small cloud of black smoke. The submersible was suddenly caught in the updraft, twisted about, and knocked into a thirty-foot-high chimney rising from the ocean floor. Dudley Foster, *Alvin*'s captain on the dive, righted the submersible and approached the chimney again, this time very carefully.

The chimney seemed to be made of some softish volcanic material. Through cracks in its surface, Foster saw it was hollow and apparently lined with crystals that glistened when the submersible's spotlight shone on them. The temperature monitor on *Alvin*'s hull began rising as the chimney was approached. When it was close enough, Foster, using his remote-controlled probe, inserted a thermometer through a crack into the core. The instrument, prepared for Galápagos dives, had an upper limit of 90 degrees. It immediately shot off scale. The crew thought the thermometer might be malfunctioning or perhaps even broken. After surfacing, they checked their equipment. The thermometer's plastic holder had melted and the rest was charred. A quick consultation of a reference book gave the melting point of the plastic as 356 degrees Fahrenheit.

The next day the increasingly excited team prepared for another dive. Bob Ballard and Jean Francheteau, coleaders of Project FAMOUS, took *Alvin* down to have a second look at the same ocean floor. They saw a series of vents, some belching out black smoke and others white smoke. Advancing cautiously toward one of them, they began taking temperature readings with a special thermometer. When inserted into the core, the gauge shot up as it had done the day before. Now there was no question of malfunctioning. Several readings reassured them that the gauges were working properly. The highest recording was 662 degrees Fahrenheit. As Francheteau later described the vents, "They seem connected to hell itself."

The discovery of these hydrothermal vents, now often simply called "Smokers," was significant. For one, it explained the abundance on the ocean floor of minerals not expected to be found there. They had been carried there from deep underground veins of ore by hot water under great pressure shooting out of the vents.

The typical Smoker begins as a jet of hot, sulfide-rich, acidic water. Porous, friable deposits of calcium sulfate, precipitating out of the seawater, begin to form a chimney around the jet, insulating it from the colder seawater. These chimneys grow about a foot a day, reaching ultimate heights of up to fifty feet. Varying in shape, with outcroppings and occasional bulbous structures, they occasionally resemble beehives. Sometimes they have one opening at the top, sometimes many, each of them typically a few inches across. The temperature can differ by 600 degrees between the tip of the opening and the cold seawater an inch away. This small inch of separation constitutes the largest natural temperature gradient on Earth.

By now Clambake has been joined by many hydrothermal

The submersible *Alvin* approaching "Godzilla," a 150-foot-high Smoker off the Juan de Fuca Ridge. *Alvin* is drawn to scale.

vent sites: Rose Garden, Rainbow, Snake Pit, Broken Spur, Lucky Strike, and Statue of Liberty, to mention a few. A hundred chimneys may exist at a 400-by-400-foot site, each having tiny metallic crystals of copper, iron, manganese, and zinc embedded in its sides. Twenty years after the discovery of the Smokers, there are commercial ventures exploring how to recover the riches. In late 1997 an Australian corporation won title to nearly 2000 square miles of the volcanic territorial waters off Papua New Guinea. Preliminary soundings revealed ore containing 15 percent copper with even more abundant zinc and considerable silver and gold. The site is estimated to contain minerals worth several billion dollars if the ore can be retrieved. That is, of course, a big "if," once again pitting mining interests against environmental ones.

The discovery of the hydrothermal vents was welcome because it tied in nicely to the notions of plate movement, of spreading ocean floors, of hot underground molten rock chambers, and of all the features of volcanic activity that go with them. The degree of activity was perhaps unexpected, but the general features were not. On the other hand, the life on those vents, both its quality and its quantity, was a total shock to the scientific community. Before the discovery of these vents, the record for thermal tolerance of a multicelled organism was an ant native to the Sahara that survived at up to 130 degrees. Here were large creatures living on the edge of environments five times that temperature. All previously known ecosystems, even on the ocean bottom, ultimately depended on photosynthesis for the production of nutrients, but the strange beings living on the vents seemed to have found another way to nourish themselves. As biologist Cindy Lee Van Dover says, "One can hardly imagine a greater contrast to the biology of the typical soft-sediment deep sea than the teeming oases of life at the hydrothermal vents."

More than 500 new species of animals have been discovered on the vents during the two decades since the first dive on the Galápagos rift. Some are similar to known species, but others are amazingly different. Bob Ballard describes the ten-foot-tall tubeworms that cluster around the ocean bottom openings this way:

> They absorb oxygen and other inorganic compounds from the water through the exposed red tips of their bodies, which look something like heads poking out of the tubes. Actually they are more like gills, with hundreds of thousands of tiny tentacles arranged on flaps lining the exposed red tip.

The tubeworms grow to their full length in less than three years and begin to reproduce after about two. They often live only a few years, sometimes dying of natural causes when the vents stop spouting, and sometimes buried by lava gushing from a split in the ocean floor. Like surface volcanic eruptions, vents have relatively short lives, perhaps a few years; the life they support ends when they end.

Although "discovered" in 1977, tubeworms are not a new species. Their scientific name is *Riftia pachyptila,* the *Riftia* part reminding us they live on rifts in the ocean floor. Fossils are often hard to identify, but in 1983, while geochemists Rachel Hayman and Randy Koski were exploring an old Arabian copper mine, the sight of a ten-foot-long worm fossil became suddenly meaningful. Copper deposits and ten-foot worms are a staple of hydrothermal vents. The conclusion was that what was now a copper mine had once been an ocean bottom. Plates can diverge under existing continents, rupturing the land on top: the Red Sea, the Gulf of Aden, and the East African rifts all developed in this kind of split. The site of the mine was known to be geologically active. So, land can become sea and sea can become land. Ninety-five million years ago, those fossils were living creatures, much like the ones *Alvin*'s crew first saw in 1977.

Another recently discovered creature, *Alvinella pompejana,* lives precariously in an inferno, moving in and out of a small tube attached to the inside wall of a Smoker. The *Alvinella* part of the scientific name reminds us it was first found using *Alvin. Pompejana* is a reference to the famous city—Pompeii—at the foot of Mount Vesuvius caught suddenly in the hot ash of the A.D. 79 eruption of the volcano. Thriving at the very edge of the hot water jet shooting up from the ocean floor, Pompeii worms, as they are commonly called, are usually encrusted with mineral grains formed from the precipitating deposits in the vents, just as Pompeii's inhabitants were covered with ash by the volcano. The proximity to the hot water jet leads to a remarkable temperature gradient along the Pompeii worm's four-inch body. Using a specially designed thermometer, Craig Carey and colleagues found that the worms' tails commonly lie in water of 170 degrees Fahrenheit, about 100 degrees hotter than their heads. The temperature gradient draws the cooler water through the tube that encases the worm, acting like a passive thermal siphon, carrying nutrients along. The worms are perfectly adapted to the hellish environment they live in.

What do *Riftia pachyptila* and *Alvinella pompejana* have in common? Certainly not size: one is ten feet tall and the other four inches long. They differ in composition as well. *Alvinella* has a fully developed digestive tract, gills, and an extensive blood circulation system, while *Riftia* has no mouth and no digestive system. However, they are alike in having adapted to live side by side, at high temperatures, on those extraordinary hydrothermal vents. They are also alike in incorporating within them bacteria varying in size, form, and function every bit as much as their hosts. It's not simply a question of scaling up or down: *Alvinella*'s filament-like bacteria are a hundred times longer than *Riftia*'s.

The bacteria are integral parts of the larger creatures that house them. *Riftia*'s very existence can only be understood within the framework of a symbiotic relationship with the bacteria that directly provide it with nourishment. Though not quite as obviously, *Alvinella* is equally dependent on the bacteria it hosts. The secret to life on the vents lies in a discovery that emerged more or less simultaneously with the discovery of the vents: bacteria with extraordinary tolerance to high temperatures. These species have come to be known as thermophiles, or heat-lovers. On vents they often coexist in a symbiotic relationship with their larger heat-loving organisms, providing nutrients and converting the toxic environment into one suited for their partners.

Some Like It Hot

The tubeworms living on the hydrothermal vents were definitely a surprise. So were the bacteria. They should all have been dead at vent temperatures. After all, that's why meat and chicken are cooked to 160 degrees Fahrenheit: to kill the bacteria. The father of modern microbiology and of the germ theory of disease, Louis Pasteur, had emphasized over and over again the importance of sterilization by heating. Proving that spoilage was caused by the arrival of microorganisms, not by their spontaneous generation, he showed how heating prevented the damage. Living by his own philosophy, "There is no such thing as pure and applied science—there is only science, and the applications of science," Pasteur wrote treatises, complete with diagrams and photographs, showing how to preserve quality by heating wine, beer, milk, cheese, or cider.

He helped design equipment to warm fluids in large quantities at low cost. His name and his work are immortalized by the common word *pasteurization.* Yet bacteria on the vents are thriving at 170 degrees or higher. What does it mean?

Thermophilic microorganisms living on Smokers have some close sea-level relatives found in hot springs. These places are similar in many ways to the ocean-bottom hydrothermal vents. Wherever water has seeped down to molten rock, it can bubble back up to the surface as a stream or a geyser. It depends on the opening. That opening may be a volcano or a peephole. The crust may have a crevice or a crack, may lie on the ocean bottom or on a mountaintop, and the manifestation may be lava or steam. When underground heat emerges on the ocean bottom or on the surface, we have what geologists call a "hot spot."

One of the most familiar "hot spots" is Yellowstone National Park's "Old Faithful" geyser in Wyoming. During the course of a long-term 1960s study of microbial life in Yellowstone's hot springs, Thomas Brock and his colleagues found a new kind of bacterium flourishing at 160 degrees. They called it *Thermus aquaticus.* Brock's group coined the term thermophile to distinguish these bacteria from the "normal" kind, with tolerance to 150 degrees set as the threshold for membership in the new and presumably rare class.

The scientists certainly weren't looking to discover many more thermophiles on land, much less on the ocean bottom. But shortly after *Thermus*'s identification, the same team found *Sulfobolus acidocaldarius* in an acidic spring at 185 degrees. To distinguish this bacterium from thermophiles, the team chose the name hyperthermophile. At that point Brock asked, What was the ultimate limit? Were there going to be superhyperthermophiles? How hot did it have to get before life became truly untenable?

The prerequisites for life anywhere in the solar system and possibly the universe are threefold. The first is an energy source capable of driving the chemical reactions that lead to life; the second is the presence of organic molecules that carry genetic information; the third is water. Liquid water is a sine qua non, the solvent and the medium for the chemical reactions. (Other liquids such as methane or ammonia have been considered from time to time, and there might even be special circumstances under which one of them works, but life as we know it needs liquid water.)

Liquid water sets unambiguous temperature limits to the internal settings of a living creature: 212 degrees Fahrenheit as a maximum and 32 degrees as a minimum. The limits are not altogether sharp because salinity lowers the freezing point of water and pressure raises the boiling point, but there is not much leeway. At temperatures close to 200 degrees, nucleic acids that make up proteins begin to lose their structure, breaking down the whole genetic machinery. This means that even without the problems caused by the boiling of water, 212 degrees becomes a reasonable upper limit.

Bacteria found on the sides of the Smokers seem to exist at the boundary between the thermally possible and thermally impossible. The present high-temperature record holder in the bacterial contests is *Pyrolobus fumarii,* which develops best at 230 degrees and will stop reproducing when the temperature dips to 195 degrees. *Pyrococcus furiosus,* not far behind, cannot grow in the presence of molecular oxygen, likes sulfur, and breeds optimally at 210 degrees. Thirty years ago, one could not have imagined such organisms and yet they exist, cannily exploiting the raising of water's boiling point past 212 degrees by the deep ocean-bottom pressure. And the surprises may not be over. Some have suggested underground hyperthermophilic life of an extent almost unimaginable, perhaps in quantities comparable to the total above-ground life. Hyperthermophiles have already expanded our views of life at the surface. Further probes, deep into the Earth's core, may expand them even more. These bacteria may hold the key to many of the secrets of life's beginnings. They may also tell us about its survival when conditions on Earth were very harsh.

On a more mundane level, thermophiles have become a rapidly growing multibillion-dollar industry, with applications ranging from complex chemical manufacturing to ordinary laundry detergents. Their particular usefulness lies in the creation of new kinds of enzymes, agents that activate and facilitate complex chemical reactions.

Ordinary enzymes often break down under extreme conditions, e.g., very high temperatures, but ones derived from thermophilic bacteria don't suffer the same fate. As an example, biotechnology demands the constant large-quantity accurate replication of a minuscule amount of DNA. This is a multiple-step procedure involving separating the DNA strands, copying them, and reassembling the finished product. The steps proceed in rapid progression at different temperatures, some of them quite high, e.g., the DNA strands separate at 200 degrees. The clockwork mechanism requires facilitators to make the whole process work—enzymes that operate at high temperatures. Presently, the favored enzyme in the rapidly growing biotechnology industry is Taq polymerase, where Taq stands for *Thermus aquaticus,* the very thermophile that started the whole field of heat-loving bacteria in the 1960s. As Pasteur taught us, yesterday's discovery is today's tool for tomorrow's discoveries.

I have focused on thermophiles because these are the best studied of the new class of bacteria that thrive under extreme conditions, but we shouldn't ignore life's more frigid temperature boundary, 32 degrees. Bacteria that live near that point, called psychrophiles, are found wherever conditions permit—in frigid oceans and in drops of water at the edges of snowfields. They eke out their existence on glacial rocks, gathering in a little sunshine, sometimes slowing their reproduction to keep pace with a colder environment. *Bacillus infernus,* living on bare rock a mile underground, divides once a year instead of the bacterial standard of once an hour, but it lives. Stretching past the 32-degree boundary, exploiting every last vantage, microbial communities exist at close to 0 degrees, but no lower. I should say "actively live" instead of "exist," since microbial communities can be preserved cryogenically at liquid nitrogen temperatures, almost 400 degrees Fahrenheit below zero. Larger organisms cannot. The freezing of water in their cells creates ice crystals that rip apart cell membranes.

Bacteria that survive under very acid conditions are another example of an interesting field of study. At the same time, enzymes derived from acidophiles are beginning to be used as food additives to improve the digestibility of cattle grain. A cow's stomach is a friendly environment for these enzymes.

For any particular organism, the temperature limit is set by that species' chosen mechanism for survival, the thermostat setting where the chemical reactions cease to work. As humans, we move up or down from our habitual 98.6 degrees. We don't function past 106 degrees and even that limit can only be sustained briefly. But it's hard to pinpoint exactly why human life stops at 106 degrees. Vascular plants can't survive past 120 degrees, but their source of life, photosynthesis, only ceases at 170 degrees. By contrast, hyperthermophiles live in the full range at which water remains liquid, 32 to 212 degrees Fahrenheit.

This raises the question of just what is the energy-generating mechanism that allows thermophiles to survive a lightless, toxic, high-temperature environment. In fact, more than survive, they even fabricate nutrients for ten-foot tubeworms. What is the secret that makes life possible on a world "connected to hell itself"? The first clue came immediately after *Alvin*'s initial dive on Clambake I. As the water samples were being analyzed, the characteristic smell of rotten eggs and sewage spread through *Lulu*, the mother ship. The samples were clearly very rich in hydrogen sulfide. Now there was a dual puzzle. How could life be so abundant on the dark ocean bottom 10,000 feet down, and how could it survive in a world containing large doses of a chemical known to kill most living organisms?

Plants make complex organic molecules out of simple ones through photosynthesis. With light as an energy source, they combine water and carbon dioxide to form oxygen and carbohydrates. Animals breathe oxygen and eat carbohydrates: life abounds. The Smoker thermophilic bacteria have found and exploited an entirely new mechanism for sustaining abundant life. They free sulfur atoms from the toxic hydrogen sulfide, combine them with carbon dioxide, oxygen, and water, and then obtain energy by reattaching the sulfur to oxygen to form a sulfate. The bacteria use that liberated energy to make carbohydrates. In essence, sunlight is being replaced as an energy source by the sulfide-to-sulfate transition. Chemosynthesis, not photosynthesis, is the new process by which life is supported on the hydrothermal vents. Furthermore, while the latter not only requires light but ceases operating at 170 degrees, the former is perfectly adapted to the temperatures found on those rifts in the ocean bottom.

This new energy source raises some very important questions, possibly affecting the future of all animal species. A great deal of discussion in the past decades has centered on what might happen were the Sun to be obscured for a long period of time. The cause could be natural, a collision with a large meteor that raises dust in the atmosphere, or a human-provoked disaster, the so-called nuclear winter. In any case, without sunlight, life would come to an end. Or would it? Could life, supported by chemosynthesis, survive on the hydrothermal vents, possibly even reemerging on the surface once sunlight shone again?

All these considerations point to thermophilic bacteria as possible maintainers of life in climatic crises, both of the future and of the past. One of the most dramatic examples of possible survival of life on hydrothermal vents occurs in the "Snowball Earth" scenario.

Snowball Earth

Life can be amazingly resilient. The biggest volcanic eruption in modern times took place in August 1883 on an island thirty miles west of Java called Krakatau. Six cubic miles of rock were ejected from the crater up into the atmosphere, and more than 30,000 people on the coasts of Java and Sumatra were killed by tidal waves, some of which traveled at a hundred miles an hour. The volcano's blast, a testimony to the thermal power inside our Earth, was heard in Australia. Dust, darkening the Sun, made the whole world cooler for five years. All that was left of Krakatau after the explosion was a small blackened nub of lava. To reflect the changes, the nearby inhabitants dropped the K and the u, reducing the isle's name to Rakata.

Nine months after the eruption, a French expedition to Rakata found a spider alive there. Five years later there were young trees, grasses, butterflies, and even lizards. The island, or rather, what the island had become, was being repopulated. However, the population wasn't the same as it had been before the eruption.

With or without such a cataclysm, species evolve through Darwin's great discovery, the mechanism of natural selection. But natural disasters act as stimulants to natural selection by reshuffling the habitat. Islands, particularly ones well separated from adjoining land, provide an especially interesting site to study evolution at work because of their relatively confined environment. The islands need not be large or even above water: the fields of Smokers at Clambake or Rose Garden are their own underwater archipelagoes sitting on mini-volcanoes. Local climate changes stimulate the emergence of mutations in a confined area and make for interesting ecological laboratories.

Foranamifera are an abundant group of microorganisms that live on the ocean bottom. They have been present in the fossil record for hundreds of millions of years. Yet suddenly, fifty-five million years ago, over half the species in the foranamifera phylum disappeared in a time period of less than 10,000 years. The evidence says they suffocated because of rising water temperatures. The reason they vanished, and this is only a possibility, reads like a detective story.

It starts well before foranamifera ever appeared on Earth and with an unlikely suspect. Single-celled methanogens—among the earliest known living forms—converted hydrogen and carbon dioxide into water and methane and helped change the atmosphere of the early Earth from one with no oxygen to one rich in oxygen. As such, they played a large role in the Earth's early climate and in the creation of the wealth of life-forms. Over the course of billions of years, the ubiquitous methanogens have produced fifteen trillion tons of methane now buried in the deep ocean floor. Around fifty-five million years ago, a sudden warming of those bottom waters, perhaps due to a shift in ocean currents, triggered the release of some of the methane into the atmosphere. Once it bubbled up, methane, with a greenhouse heating capacity thirty times as high as carbon dioxide, produced further warming.

Perhaps the temperature didn't or couldn't readjust quickly enough by a feedback loop. The temperature kept going up. This caused more release of methane, triggering a runaway greenhouse effect. During the next 10,000 years, a trillion tons of methane, 7 percent of the ocean floor total, made its way to the atmosphere. At that point the Earth finally managed to reequilibrate. But by then the ocean temperature had risen 10 degrees. Half of the foranamifera species, survivors of many millions of years of climate change, perished because they were unable to adapt to the higher temperatures. The fossil corpses lie in jumbled piles of mud formed by gas, probably methane, bubbling up from the ocean bottom. In shorthand, the story is sometimes told as "The forams were killed by a methane burp." Thus, sudden climate changes are likely to lead to new paths in evolution's march. The new species are the favored ones.

Temperature affects all forms of life. Some thrive in the cold, some in the heat. Some survive climate changes, some not. Many of the known modern mammals, from rodents to primates, made their first appearance in the fossil record fifty-five million years ago. Animals with hooves, claws, and stabbing canines appeared. Teeth changed, hunting patterns altered.

A runaway greenhouse effect that comes under control after 10,000 years and a 10-degree warming of the oceans is dramatic. Now consider the possibility of a 200-degree global temperature jump in one century. Sounds utterly absurd and yet, if Paul Hoffman and Daniel Schrag are correct, it happened over half a billion years ago at least once, and perhaps several times. Hoffman and Schrag call these jumps "freeze-fry" events. For them, they are a key part of the "Snowball Earth" scenario.

The period between 750 million and 550 million years ago is puzzling geologically and biologically. Stones of that period, with the characteristic striations marking them as glacial detritus, are found in the wrong place, near the equator. This is a familiar story: Agassiz and then Lyell and Buckland had used similar evidence to infer earlier glacial presence in Switzerland, France, and Britain. In this case the stones suggest ice once covered the whole Earth. On the other hand, layers of carbonate rock produced by precipitation from warm carbon dioxide–laden water lie right on top of glacier remains. Apparently the Earth went quickly from being very cold to very hot. Moreover, there are layers of iron-rich rock, suggesting lack of oxygen. But oxygen, thanks in no small measure to the methanogens, had been abundant on Earth for a billion years.

Then there is the question of life. It emerged on Earth about 3.8 billion years ago as simple single-celled creatures. The fossil record is hard to interpret, but evolution was undoubtedly slow for the next three billion years. Suddenly, about 565 million years ago, new organisms surfaced in profusion. Within a space of fifty million years, a blink of the eye on these time scales, representative members of all the basic forms of animal life appear.

Phylum is the Greek word for "tribe." By 500 million years ago, the eleven known animal phyla—the worms, the starfish, the mollusks, and all the others, including the chordates, our own tribe—were thriving. Furthermore, no new phyla have appeared since then. This sudden burst was singular. There have been massive destructions and births of species, but none comparable in magnitude to the proliferation of life in the "Cambrian Explosion."

What is the connection, if any, between the sudden changes of global temperature, the rapid shifts up and down of oxygen and carbon dioxide, and the explosion of a variety of life-forms? A possible answer is emerging. It's not universally accepted, there are loose ends, but it holds together remarkably well. The story hinges on a series of ups and downs of the Earth's temperature, all of them caused by factors we've already encountered: Agassiz's glacier movements, Croll's feedback loops, a runaway greenhouse effect, Wegener's continental drift, and thermophilic bacteria. Each of them is crucial at some point in the tale's unfolding. It's a tale whose contributors come from around the world, each one providing another piece of the puzzle. From Cambridge, England, to St. Petersburg, Russia, to Pasadena, California, to Cambridge, Massachusetts. It's characteristic of the give-and-take of modern science.

In the early 1960s, the resurrection of Wegener's theory of continental drift persuaded W. Brian Harland that the continents had clustered around the equator 700 million years ago. Since land surfaces reflect back less sunlight than snow or ice, but more than open water, equatorial regions were cooler than they are in the present ocean-dominated configuration. That's step one: a cooler equator. Using early global climate models, Michael Budyko showed a cooler equator would trigger a feedback loop in which polar glaciers would start spreading southward, amplifying the equatorial cooling. Step two—a runaway freeze mode. This ends with the whole Earth covered by a miles-thick layer of ice. There was no step three because it wasn't clear how to melt the ice or how life could be preserved through a big freeze.

That's where matters stood until the early 1990s, when Joseph Kirschvink, the originator of the term "Snowball Earth," pointed out that large amounts of carbon dioxide would be released into the atmosphere as a result of volcanic bursts. These eruptions, unperturbed in their creation by happenings on Earth's surface, would be strong enough to penetrate the icy covering. Carbon dioxide is normally kept in check through absorption from plants as well as by burial in rocks washed into the oceans after erosion and exposure to rainfall. It's a delicate balance. Six hundred million years ago, with plant life frozen and rocks and oceans covered by ice, there was no check to the carbon dioxide buildup. The stage was set for step three: a runaway greenhouse-gas-induced warming.

A few million years of volcanism without carbon dioxide–absorbing mechanisms may have led to carbon dioxide levels in the atmosphere of 120,000 parts per million, over 300 times as high as they are now. Step three kicked in dramatically. Within perhaps as little as a century, the average global temperature rose from a frigid 60 below zero to a steamy 120 above. On a Celsius scale, the shift takes a particularly symmetric form: –50 to +50, a 100-degree turnabout. As this occurs, the ice melts and the Earth switches from glacial to tropical. With regard to the disappearance of oxygen, that's simple. The freeze had killed plant life, the source of oxygen. The geological record is explained. In sequence, we have glaciers, no oxygen, and finally carbonate rock from warm, carbon dioxide–rich rain.

As to how life survived on an ice-covered Earth, here's where the phenomenon of thermophiles and their hardy brethren may have saved the day. In a deep freeze, life could have continued on isolated underwater hydrothermal vents while it died on the surface. The stress on evolution acted as a bottleneck: most if not all of the primitive microorganism species on the surface might have died out. As the Earth thawed, surface life returned, generated in new and more varied forms, soon to multiply in astounding profusion. Perhaps the deep freeze was even a trigger of sorts for that spurt.

"Snowball Earth" may not be entirely right, but many

scientists are moving toward acceptance, some with modifications. William Hyde et al. recently proposed a partially frozen Earth, with ice-free oceans near the equator and carbon dioxide levels four times, not 300 times, as great as present ones. They aren't as confident as Hoffman and Schrag about life's ability to survive a deep freeze. In a recent exchange in *Nature,* the latter say that "Snowball Earth" fits the data while "Slushball Earth" doesn't. Hyde responds:

> Further data may call for the reassessment of our open-water scenario, but we consider that the hard–snowball earth scenario is not yet proven. We believe that the open-water solution is much more favorable for the survival of metazoans, allowing their remote progeny to continue this discussion.

And so we, remote progeny of metazoans, the earliest multicellular organisms, shall continue the debate and very likely succeed in settling it during the next decade. No part of the "Snowball Earth" argument is unfamiliar. There is abundant evidence for past ice ages; this one was just more extreme. Venus's greenhouse gas warming is a lot bigger; it didn't stop until the temperature reached 800 degrees. By comparison, 120 degrees is moderate. Finally, thermophiles have surprised us more than once and will probably continue to do so.

One of the biggest surprises, though we didn't know it at the time, came in the same year *Alvin* was diving down to Clambake I. In 1977 a University of Illinois microbiologist named Carl Woese discovered that methanogens, contrary to everyone's expectations, are not bacteria. Many of them are a new kind of thermophile. They provide the key to a set of revelations that have reshaped our thinking about life's origins and the Earth's temperature when those first organisms appeared.

The Third Branch of Life

Until the late 1970s, the standard scientific view about the origin of life had it starting in a pool on the Earth's surface with the production of simple organic molecules. In a classic 1953 experiment, chemists Stanley Miller and Harold Urey filled a largish flask, a little bigger than a gallon, with water, methane, hydrogen, and ammonia, all supposedly present on the early Earth. They then sealed the flask. Exposure to electrical discharges and circulation through a heater and condenser simulated the conditions that a mixture like it would have met: lightning and temperature changes. Organic molecules were found in the mixture after only a few days; these were hailed as the precursors of life.

Despite this attractive simplicity, a number of questions remain that challenge this view of life's origins. Although the presence of organic molecules is suggestive, there's no obvious mechanism for their assembly into the larger RNA and DNA molecules that are the true carriers of a living organism's genetic information. The abundance of methane, hydrogen, and ammonia that the Miller-Urey experiment required may not have been present when life began, and the surface of the early Earth probably looked far less benign, thermally and otherwise. With that theory challenged and promising new constructs emerging in the late 1970s, scientists began thinking about a possible origin on hydrothermal vents.

One of the first to do so was Jack Corliss, the scientist who stared in disbelief out of *Alvin*'s porthole during the initial dive over Clambake I. Stunned by all the life he saw,

Corliss asked, speaking by acoustic telephone to a graduate student on the support ship *Lulu,* "Isn't the deep ocean supposed to be like a desert?" Seeing the abundant life now clustered around the vents made him think this might be the place where life started. Realizing that the ocean floor is relatively protected from collisions of the Earth with asteroids, he thought it would not be as vulnerable to sudden changes in sunlight and temperature as the surface. In 1981 Corliss and two collaborators published a paper entitled "Hypothesis Concerning the Relationship Between Submarine Hot Springs and the Origin of Life on Earth."

At about the same time, the classification of living organisms underwent a major change, one that lent support to the notion of life originating on "submarine hot springs." Prior to the late 1970s, the division had been between bacteria, also known as prokaryotes, single-celled organisms whose cells did not contain nuclei, and all other forms of life. Those others, eukaryotes, have cells with true nuclei, whereas bacteria lack them. Supposedly a later arrival, the eukaryotes, when characterized by form, function, and size, are divided into four kingdoms: animals, plants, fungi, and algae.

In the mid-1960s Carl Woese, a microbiologist by training, started to provide ordering in the bacterial world. This was a considerable challenge at a time when the leading bacteriology textbook said, "The ultimate scientific goal of biological classification cannot be achieved in the case of bacteria." Rejecting this thesis, Woese took it simply to mean that studying under a microscope the form, function, and size of bacteria is insufficient; he thought the emerging techniques of molecular biology provided a new means of attacking the problem.

In using new tools, Woese focused on a particular microscopic kind of RNA known to reside on the intracellular sites where proteins were assembled. By looking at the degree of overlap in genetic material, he could trace bacterial lineage, the time when two species diverged. In effect, he was creating a universal bacterial chronometer.

In 1976 a colleague of Woese's suggested looking at methanogens, the very same organisms that led indirectly to the death of the foranamifera. Woese was immediately intrigued because methanogens have a variety of shapes: round, rods, or spirals. They also have a variety of sizes. However, despite the differences in appearance, all methanogens perform the same chemical reaction. Function was the same, but not form or size. If their RNA was as similar as Woese thought it would be, they could provide a powerful confirmation of his new way of classifying bacteria.

Then came the bombshell. Methanogens' genetic sequencing seemed not to be bacterial. Initially cautious, Woese labeled them archaeabacteria. Most biologists dismissed his work. As one researcher put it, he was "a crank, who was using a crazy technique to answer an impossible question." Supporting evidence began to accumulate as Woese and other scientists studied methanogens' molecular details. The "crank" was turning into a hero. Fairly soon, the added "bacteria" was dropped and archaeabacteria became known simply as Archaea, a third form of life, distinct from both eukaryotes and prokaryotes.

A dramatic confirmation of the correctness of Woese's point of view has been provided by modern genetic techniques. The methanogen *Methanococcus jannaschi* is one of the first organisms to have its DNA fully sequenced. Forty-four percent of its genes resemble those of either bacteria or eukaryotes, but the rest are completely different. Unequivocally, it's "something else." *Methanococcus jannaschi* has another significant feature. It's a hyperthermophile, growing optimally at close to 180 degrees Fahrenheit. Almost all the Archaea discovered in the first few years were thermophiles, hyperthermophiles, or "lovers" of some extreme or other—pressure, salt, acid. The whole group came to be called extremophiles, with thermophiles as a subset. *Pyrolobus fumarii,* found on Smokers, still the temperature tolerance record holder, isn't a bacterium. It belongs to Archaea.

The closeness of Archaea to the roots of the tree of life and the evidence of their extremophilic nature suggest they were the first microorganisms. Unfortunately, evolutionary biology is seldom crystal clear. Archaeans are far more abundant than we originally thought. They are present in marine plankton, in temperate waters, and one kind is even found living in polar waters. Also, all hyperthermophiles are not Archaea. *Aquifex aeolicus* lives at 200 degrees, but is a bacterium. Perhaps the first microorganisms were thermophilic, but neither simply Archaean nor simply bacterial. It may even be wrong to think of life having a single root. As Carl Woese recently said,

> The ancestor cannot have been a particular organism, a single organismal lineage. It was communal, a loosely knit, diverse conglomeration of primitive cells that evolved as a unit, and it eventually developed to a stage where it broke into several distinct communities, which in turn became the three primary lines of descent [bacteria, archaea, and eukaryotes].

The discovery of Archaea and their close link to thermophiles has not answered the how, when, and where questions about life's origins, but it has changed the discussion's parameters. Much of the difficulty in pinpointing the answers, if that's even possible, has a common source: scientists are united in fixing the approximate date and cause of our Earth's origin but uncertain about its temperature during its first few hundred million years.

Melting the Earth

A temperature gradient shaped our solar system's eight planets: the four inner ones are small, dense, and rocky and the outer four are large and gaseous. Temperature determined each planet's size and composition at birth and continues to influence their evolution. Temperature's ups and downs have shaped and reshaped Earth's surface, often destroying life and just as often stimulating its rebirth. To understand how and why that happened, we need to go back in time five billion years, to the formation of our solar system.

Around then a large nearby star died in a spectacular explosion. That explosion, or rather the shock waves it produced, created ripples in the nearby interstellar medium. These are regions of higher and lower density, higher and lower temperature. One of them, let's call it a cloud, was perhaps a little denser and hotter than most. It may also have been set spinning by the shock wave. As it spun, the cloud flattened into a whirling disc with a central bulge. The disc was mainly composed of hydrogen and helium, the primitive and basic constituents of the universe. The disc also contained grains of other materials, formed in the core of the dead star and shot outward in the star's explosive end.

Gravity continued to draw the disc's bulge inward, driving pressure and temperature upward. A protostar, the precursor to a star, was formed as the temperature rose. Later, when the core of the bulge reached fifteen million degrees Kelvin, nuclear reactions began and our Sun emerged, a new full-fledged star. The whole process is not unusual. To see it happening now, simply look for evidence in the spiral arms of our Milky Way.

While the bulge was becoming a protostar, the heat it liberated was shaping the disc's outlying dust and grains. Temperature dropped going away from the center. In the disc's inner part, it was over 1000 degrees Kelvin at Mercury's present location and 400 degrees Kelvin at Mars's. Over the course of a hundred million years or so, in a scenario confirmed by computer simulations, the rocky grains and dust grew into rocks and eventually formed four planets: Mercury, Venus, Earth, and Mars.

The outer regions of the disc have a different history. Since the temperature there was so low that vapors froze, the grains acted as nuclei in the formation of giant ice-balls. Within a few million years these balls became several times bigger than the Earth. Composed of frozen methane, ammonia, and water ice mixed with rocks, they acquired enough mass to attract hydrogen and helium out of the disc and grow further. Eventually they too stabilized into four planets: Jupiter, Saturn, Uranus, and Neptune.

The eight planets fall nicely in place in this rotating disc picture. Each of the inner four, the terrestrial planets, is about five times as dense as each of the outer four, the Jovian planets. Ultimately, the difference in size and density between the inner four and the outer four is due to the temperature at which they were born. As for Pluto, a bit of an outcast, it may not have been formed this way and perhaps shouldn't even be called a planet.

Many pieces of evidence support this picture of inner four and outer four planets. One, known for hundreds of years, puzzled Newton. Whereas comets move in random directions, the planets revolve around the Sun in almost circular orbits, all lying in approximately the same plane and all turn-

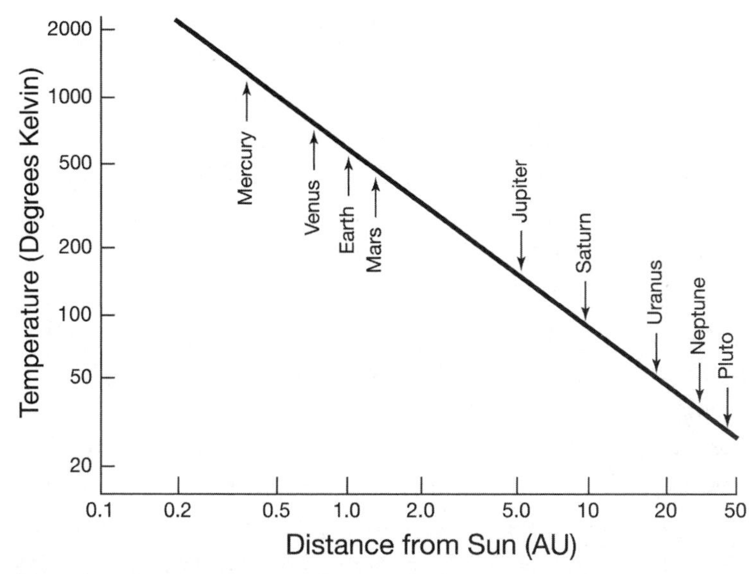

Plot of temperature versus distance in the solar system at the time planets were formed (unit of length is the astronomical unit equal to the Sun-Earth distance; vertical axis not drawn to scale)

ing about the Sun in the same counterclockwise direction. To Newton, who had explained the basic laws regarding planetary motion, this placement in a plane seemed inexplicable by the laws of science, leaving him no alternative but to invoke a higher authority. In a 1692 letter to his friend Dr. Bentley, Newton wrote, "the motions which the planets now have could not spring from any natural cause alone, but were impressed by an intelligent Agent." Over a century later the "natural cause," birth in a whirling disc, was identified by the French mathematician Pierre Simon de Laplace and by the German philosopher Immanuel Kant.

In the early solar system, a little more than 4.5 billion years ago, some of the original debris from violent collisions eventually made its way into the eight (or nine) planets. Some simply stayed as dust. Other parts grew to form asteroids, rocks ranging in size up to a few hundred miles across. Now mostly clustered in a belt between Mars and Jupiter, 200 of them are bigger than fifty miles across and 100,000 are large enough to show up on photographic plates. Smaller chunks of rock, less than a few hundred feet across, are usually called meteors, though the dividing line between asteroids and meteors is often blurred.

Billions of years of planetary stability has settled asteroid paths into regular orbits of their own, occasionally changed by a perturbation of one sort or another. In that case, a good-sized rock may be set on a collision course with our Earth, an event that has happened many times in the past few billion years. Despite its frequency, the idea of rocks falling from the sky seemed strange as late as the nineteenth century. Thomas Jefferson, after hearing a lecture by two Yale professors on the subject, is said to have remarked, "I could more easily believe that two Yankee professors could lie than that stones could fall from Heaven."

About 300 tons of extraterrestrial rock, most of it in the form of dust, rains on the Earth every day. Occasionally large "rocks" collide with our planet. An iron-rich meteor a hundred feet across struck the ground near Winslow, Arizona, 50,000 years ago. Traveling at 25,000 miles an hour, the meteor left a crater 500 feet deep and almost a mile across. Comparable to a large hydrogen bomb, it blew enough dust into the air to affect global climate. We can expect an impact like it every hundred thousand years. Every few million years the Earth can expect an even bigger collision and every hundred million years a still bigger one.

The most dramatic collision of the past hundred million years is one first established in 1980 by the father-son team of Luis and Walter Alvarez. Luis was one of the century's great experimental physicists, a Nobel Prize winner and a man noted for being interested in many areas of science. Luis's son Walter, a geologist, was intrigued by the presence of a clay layer that appears in several places on Earth as a divider between two significant geological eras. The clay layer, little more than an inch thick, is also notable because the time of its deposition, sixty-five million years ago, marks a significant transition in life forms. In that transition, almost three-quarters of known marine species disappeared and all of the dinosaur species vanished, never to reappear. Before the Alvarezes' work, there were many hypotheses of what might have caused the extinction: change of vegetation, volcanic activity, and even asteroid collisions. None was altogether convincing.

Walter Alvarez began by asking himself how long it had taken for that clay layer to be deposited. His father had a suggestion of how to estimate the time interval: measure levels of a rare element. Iridium is normally very scarce on the surface of the Earth because almost all of the iridium that was part of the Earth's original makeup sank to the core during our planet's early molten history. By comparison, although far from asteroids' major constituent, there is often a fair amount of iridium scattered throughout them. Asteroid dust falls on the Earth at a more or less constant rate, 300 tons a day, of which a small but measurable quantity is iridium. Since the dust is uniformly distributed and essentially all the iridium on the Earth's surface comes from asteroids, you know how much of it lands each day on every patch of Earth. A measurement of the clay's iridium content would tell how many days it had taken for the clay layer to form.

Excellent approach, thought Walter, who collected the clay and sent it off for measurements. All was well until the results came in: the clay had more than a hundred times the expected amount of iridium. Either the clay had taken a hundred times longer to form than he thought was possible, or something else had happened sixty-five million years ago, something that deposited lots of iridium on Earth at that time. This was more than just daily dust. The Alvarezes thought the cause might be an asteroid collision with the Earth while the clay layer was being formed. If the asteroid weighed enough, its iridium contribution could be much bigger than that expected by centuries of dust. This certainly fit the data. However, the Alvarezes couldn't initially see how a large impact would cause worldwide extinction. As Walter Alvarez writes,

> Finally Dad started thinking about the dust that would be thrown into the air by an impact. He remembered reading that the 1883 explosion of the Indonesian volcano, Krakatoa, had blown so much dust and ash into the atmosphere that brightly colored sunsets were seen for months in London, on the other side of the world, and he tracked down the book he remembered. Scale the Krakatoa event up to the size of a giant impact, thought Dad, and there would be so much dust in the air that it would get dark all around the world.

That was it. A very large asteroid, a 1000-billion-ton rock, collided with Earth and shot up enough dust into the atmosphere. That was enough dust to obscure the Sun for years and to lower the surface temperature of the Earth. During the "nuclear winter" that followed the asteroid collision, all the species of dinosaurs died, unable to cope with the temperature change. Only smaller, more versatile, and more adaptable species survived. The Alvarezes estimated that the asteroid was about six miles across with an impact equivalent to the simultaneous explosion of seventy million hydrogen bombs. Luis Alvarez would know; he was an expert at measuring big explosions. He had been at Los Alamos and was the physicist in charge of measurements when the *Enola Gay* flew with the first atomic bomb.

Originally there was a good deal of skepticism about the Alvarezes' scenario, in large part because there was no known crater from that period. In 1992 a group of geologists spotted one centered in the Mexican waters just off the Yucatan peninsula, near Chicxulub. It hadn't been noticed previously because it was covered by sediment. The glassy debris in the crater and the shock patterns in the surrounding rock suggested it had been formed by a very large high-temperature impact. Its width was about 120 miles, consistent with shaping by a six-mile-wide asteroid traveling at 25,000 miles an hour. At those speeds, the asteroid crossed the atmosphere in a second, compressing the air in front of it. It plowed through mud, through two miles of limestone, and entered the granite layer below. Compression drove temperatures up to 40,000 degrees, more than enough to produce the fused quartz in the crater. The dating of the rock gave an estimated crater age of 64.98 million years. By now, there is little doubt that the dinosaur extinction was caused by an asteroid-induced temperature change.

This made scientists look harder at other extinctions. About 250 million years ago, the mother of all mass extinctions took place. Eighty-five percent of all the marine species and 70 percent of land vertebrates that then existed disappeared. Very recent evidence shows that an asteroid also caused much of that disappearance in an extinction numbered in tens or hundreds of years, not millions, as was once thought. A seventy-five-mile-wide crater was recently found in the Australian outback, near Shark Bay on the Indian

Ocean. The date of the rock is consistent with a collision 250 million years ago and the crater has fused quartz in it, suggesting high temperatures caused by a high-velocity impact. Although there aren't large amounts of iridium, chemical analysis of the rocks shows they contain helium and argon of the types one expects in asteroids. Normally the gases would have floated away, but they were trapped inside buckyballs, the geodesic-dome-like molecules formed by sixty atoms of carbon at high temperature. The three pieces of the puzzle come together again: the extinction, the crater, and the chemical analysis.

Dinosaurs emerged after one collision extinguished the early reptiles and disappeared after a second collision. One temperature shift stimulated their evolution and a second one, 185 million years later, caused their extinction. Jefferson's two "Yankee professors" were right: rocks, even big rocks, fell on the Earth. They destroyed old forms of life and created the conditions under which new forms would emerge. Perhaps one of them even carried the first life to Earth.

Extraterrestrial Life

Early in our Earth's history, before asteroids settled into their regular orbits, impacts like the one that killed the dinosaurs were very frequent. The biggest collision of them all occurred 4.5 billion years ago. Our planet, then in the late stages of its own formation, was struck a glancing blow by an asteroid not much smaller than the Earth itself. The collision shaved off a large slice of Earth's surface, knocking it into space. Much of the debris, liquefied by the collision if not already molten, entered into orbit, was cooled, and then reconstituted as our Moon. Sounds far-fetched, but lunar rocks recovered by *Apollo* astronauts confirm the picture. Known as "the big splat," the impact tilted the Earth relative to its plane of rotation. The "big splat" produced our Moon, caused the seasons by tilting the Earth, and contributed to the regular alternation of nights and days by affecting the Earth's spin.

A tremendous amount of energy was released during the repeated early bombardment, much of it retained in the Earth as heat. That stored heat and natural radioactivity remain the two principal sources of energy within our Earth, the driving force of volcanism, plate movement, and all the other thermal activities that still stir up our planet. Direct surface evidence of those collisions has been lost; we can't see any impact craters from that time because heat has reshaped our planet's surface. In contrast, the pockmarked exteriors of Mercury, Mars, and our own Moon bear witness to early collisions. Venus's surface, reshaped by heat like ours, does not.

The temperature of the Earth during its first 600 million to 700 million years is the key to determining when and where life started. The earliest known fossils date back to approximately 700 million years after the formation of the Earth and the Moon. Most students of evolution regard that as a comfortable interval for life to appear. However, if we find the Earth hadn't cooled enough for life to form until, say, 3.95 billion years ago, and yet fossils were already present 3.85 billion years ago, the picture changes. One might well be skeptical about whether a hundred million years is long enough for simple organic molecules to have found, by trial and error, how to assemble themselves into the complex molecules that regulate life's genetic machinery.

If there wasn't enough time, life must have reached Earth, already formed, from somewhere else. At present we don't know how much is enough, the crux of the dilemma. Keeping in mind that there is no clear and firm answer to when or how life started, let's look at the various pieces of evidence we get from reading temperature. Consider the big collision 4.5 billion years ago. It generated enough heat to raise the surface temperatures to well over 200 degrees, enough to vaporize any water present. The secondary effects of the collision are less clear. On the one hand the release of carbon dioxide from pulverized rocks produced a huge greenhouse effect, raising surface temperatures, but the dust also created a 2000-year Sun-obscuring and cooling "nuclear winter." Determining which of these effects was more important in setting temperature for both the short and the long run requires detailed calculations. The Earth's surface could have been frozen for long periods, punctuated by asteroid-induced infernos during which temperatures shot up and the surface dissolved.

Early thermal history is so hard to determine because surface melting wiped the slate clean, erasing prior records. The oldest known rocks on Earth are about four billion years old, formed when the worst of the meteor collisions was finally over. But the oldest grains of rock, microscopic zircon crystals, are 4.4 billion years old. Recent chemical analyses of their composition suggest they were formed in the presence of liquid water, perhaps water formed in the collision with a water-ice-rich comet. That doesn't mean life did start then, but some forms of life might conceivably have arisen that early. As for survival during subsequent climate changes, the extraordinary hardiness of extremophiles lends credibility to this extreme possibility.

I would like to underline that the beginning of life four billion years ago and the emergence of phyla in the Cambrian Explosion 600 million years ago are not related to one another in any way. They do, however, share some superficial similarities: in both cases, remarkable changes in the organization of life took place in relatively short time spans while the Earth was undergoing dramatic shifts in temperature and other climate features. Just how those rapid shifts stimulate change is far from clear. In the first case, as I have already emphasized, it's not even known if life did originate on Earth.

Mars may have been a life-friendlier environment in the 700-million-plus-year interval between the formation of the solar system, 4.6 billion years ago, and the first known appearance of life on Earth. The now extinct volcanic activity on Mars produced enough carbon dioxide to give a sizable greenhouse effect and liquefy whatever water was present there. If life did originate on Mars, how did it get to Earth and how did it survive the journey? It's known that material from Mars still reaches us, so the chances are high that the trip was also made when asteroid collisions were more violent and more frequent. As for survival, here's where extremophiles come in again. The voyage from Mars to Earth is not an easy one, but some very hardy forms of life might make it, tucked away in a rock crevice, shielded from solar ultraviolet radiation.

Interest in extraterrestrial life was piqued by an August 7, 1996, press conference that NASA held in Washington, D.C., to announce the identification of Martian meteorite ALH84001, a four-pound rock found in Antarctica. That's the best place to find meteorites because they stand out so clearly against the bright white surface of the slowly accumulating glaciers. The gradual flow of ice tends to make them pile up where a glacier runs up against a mountain, providing a natural hunting ground for meteorites. Thousands of them have already been found in Antarctica's Alan Hills, hence ALH, a favorite location. This rock, however, looked a little different from the others, a little older if nothing else. Its history has now been completely reconstructed. Crystallized from molten rock 4.5 billion years ago, it underwent a thermal shock 500 million years later, perhaps when Mars was hit by a large asteroid. Water flowed through crevices of ALH84001 a little over three billion years ago. Knocked off Mars's surface sixteen million years ago, possibly by a glancing meteor collision, it landed on Earth 13,000 years ago. The trip was only half the problem.

Meteorites can easily heat up to more than 1000 degrees as they plunge through the atmosphere. We see them as "shooting stars" on a summer night. Fortunately the heat doesn't penetrate deeply; unless the rock breaks up, the inner parts survive the temperature shock. ALH84001 was particularly exciting, hence the press conference, because of sausage-shaped structures in it resembling Earth's smallest bacteria. The rock also contained chemical compounds often bacterially produced. Is this evidence that life once existed on Mars?

The thought of extraterrestrial life reaching us is fertile ground for science fiction writing, much of it unencumbered by the basic rules of biology, chemistry, and physics. However, many customarily strict adherents to those rules are willing to stretch them in considering space travel. Svante Arrhenius, the physical chemist encountered in the last chapter as the first calculator of global warming's magnitude, began thinking more than a century ago of life arriving from elsewhere. He envisaged thermally resistant microorganisms gently floating everywhere in space, sprouting where they settled. Arrhenhius coined the term *panspermia*, or "seeds everywhere," to describe the phenomenon.

In this century the notion of panspermia has been put forth again, most forcefully by Francis Crick and Leslie Orgel, two of the greats of modern molecular biology. Their narrative lodges the seeds in an unmanned spaceship sent by a civilization located outside our solar system. The spaceship is designed to protect the seeds against damage on the long trip. On arrival, the ship drops seeds in the deep Earth ocean. The theory, named Directed Panspermia, has its interesting features, but as Crick himself says, "The kindest thing to state about Directed Panspermia, then, is to concede that it is indeed a valid scientific theory, but that as a theory it is premature. This inevitably leads to the question, will its time ever come? And here we must tread cautiously."

We should indeed tread cautiously in assessing the chances of life existing elsewhere in our own solar system, independently of whether or not the two forms of life have a common source or ever cross-pollinated. Most scientists think the chances of independent evolution are higher than those of any form of panspermia, but there is no substitute for hard evidence. The conditions on early Mars were once favorable and Mars still remains a candidate despite its present frigidity. The moons of Jupiter may be a better bet.

Life Under Two Miles of Ice

Mars and the outer planets are too cold, too far away from the Sun for life above ground. But we know from the creatures found on the Earth's hydrothermal vents that life can exist even in the absence of sunlight provided an energy source is present and the temperature is high enough for water to remain liquid.

The extraterrestrial candidate most likely to harbor life is an unlikely one: Europa, one of the moons of Jupiter. As for Jupiter's other moons, Callisto appears to have water, but little thermal activity while Ganymede has a great deal of the latter, but little or no water or at least was thought to have little water until very recently. For that matter, we shouldn't altogether ignore Io, the fourth of the moons. Its outrageously high volcanic activity dwarfs that of anything we know, with huge calderas spilling 4000-degree-Fahrenheit lava onto the surface, hotter than terrestrial volcanoes have been for three billion years. An interesting phenomenon, one that may even shed light on our own planet's distant past, but Io doesn't seem to present a life-friendly environment. In short, the moons of Jupiter show promise for spawning life of one sort or another, but no clear presence.

When Galileo discovered the four moons of Jupiter in 1610, he was amazed. They confirmed for him the Copernican notion that the Earth revolved around the Sun. If Calypso, Europa, Ganymede, and Io circled Jupiter, the Earth could certainly circle the Sun. Wouldn't it be ironic if the same four moons that showed we are not the center of the universe also showed Earth is not the center of life? The symmetry would have been particularly apt if the discovery had been made by the *Galileo* spacecraft, named to honor the first observer of the Jovian moons, but symmetry is seldom so perfect. The spacecraft, equipped with a wide range of detectors, recently completed a four-year tour in which it orbited Jupiter twenty times. It looked carefully at the four big moons, its closest approaches being a little over a hundred miles away.

One of the big questions, spurred by our knowledge of extremophiles, is whether Jupiter's moons can support life. The *Galileo* probe focused on Europa, whose icy crust has an average surface temperature of –260 degrees Fahrenheit near its equator and almost a hundred degrees colder at the poles. The spacecraft's pictures reveal deep cracks in that surface, suggesting a ground torn by upwelling thermal activity. There are visible dark ridges where briny mineral fluids may be breaking through. Elsewhere, Europa's exterior shows jagged mounds that resemble icebergs in a slushy Arctic sea; again hints of a thermally active interior. The suspicion is corroborated by the absence of comet- or meteor-induced impact craters, suggesting that underground thermal activity recently reshaped the surface. All these clues point to heat generated in Europa's core, but is it enough to maintain a deep ocean of liquid water? Probably yes, but we won't know for sure until a spacecraft detects the sloshing back and forth of water, tides in an underground ocean.

The Jovian quartet and Mars aren't the only candidates for extraterrestrial life in our solar system. Saturn's Titan moon, Neptune's Triton, and even Pluto's Charon are all possibilities. Planets surrounding other stars have been discovered in the last decade. If these stars are like our Sun, and many of them are, why couldn't some of them or possibly their moons carry life? Freeman Dyson says one or more of the numerous asteroids between Mars and Jupiter might also be contestants.

The answer to many of these puzzles will come in the next decades as we probe nearby planets and their moons with more precise space missions. Interestingly enough, even as our horizons expand and we look outward, important data will likely come from our own Earth. That information is in the bottom of a lake at a place where nobody would expect to find lakes.

Russian scientists have staffed a research station on the Antarctic shelf at Vostok for almost forty years. Close to the South Pole, it's widely believed to be the harshest place humans have ever lived: recently the temperature at Vostok fell to 132 degrees Fahrenheit below zero, the lowest ever recorded on Earth. Base scientists have to deal with the dryness, the remoteness, the darkness, and the cold. In addition to all that, the base is at 12,000 feet, making it almost unlivable for even the hardiest. Vostok residents sleep with oxygen bottles by their side and almost all take medication for altitude sickness.

Researchers not only stay at Vostok. They work. In a

monumental effort, teams have slowly and steadily drilled downward through 400,000 years of accumulated ice. The great dryness of the air, comparable to that of the Sahara, leads to so little snowfall that the ice at Vostok is less than four kilometers or 2.5 miles deep. The original purpose of the drilling was to study climate change. Like rings in a tree, the ice layers maintain a history of temperature changes, in this case 400,000 years long.

Amazingly enough, water has been discovered beneath the ice. The first indications of an underground lake came in the 1970s from radar blips of overflying aircraft, but the lake's true extent only became clear after the European Remote Sensing Satellite's 1996 analysis. Lake Vostok is a vast underground reservoir, as big as Lake Ontario and, at 2000 feet, deeper than Lake Tahoe. Sitting on a rift in the tectonic plates similar to the one under Siberia's Lake Baikal, Lake Vostok is a prime candidate for hydrothermal vents.

This combination of geothermal heating and liquid water under miles of ice is the reason why astrobiologists are so interested in Lake Vostok. If there is anywhere on Earth likely to resemble the frozen layers of the Jovian moons, this is the place. By now Vostok is an international research center. A joint Russian-French-U.S. team drilled to within 400 feet of the lake's surface in 1998. Why did they stop? Oceanographer David Karl put it succinctly: "No one wants on their tombstone 'I polluted Lake Vostok.' "

The last 300 feet were drilled through refrozen lake-top water. The sample indicates the lake has inorganic nutrients, dissolved carbon, and bacteria, all the ingredients of an active ecosystem. The researchers realized they were about to enter a vast pristine environment that had evolved in isolation for close to a million years. Once that ice-lake border is breached, any outside contaminants will forever bring into question the results of subsequent measurements. This consideration resulted in repeated calls for maximum care. So they stopped drilling and, until the necessary precautions can be taken, they will not resume. It's a difficult situation. The drilled hole is kept open by sixty tons of a mixture of aviation fuel and Freon, one ton of which lies in ice refrozen from lake water. It's too much to be hauled up and stored; yet, if it is allowed to enter the lake, it will drift to the bottom in days and within years circulate through the rest of the lake.

The Vostok drilling has taught us a great deal about climate and temperature's ups and downs over the past 400,000 years. A new instrument is needed to carefully teach us about

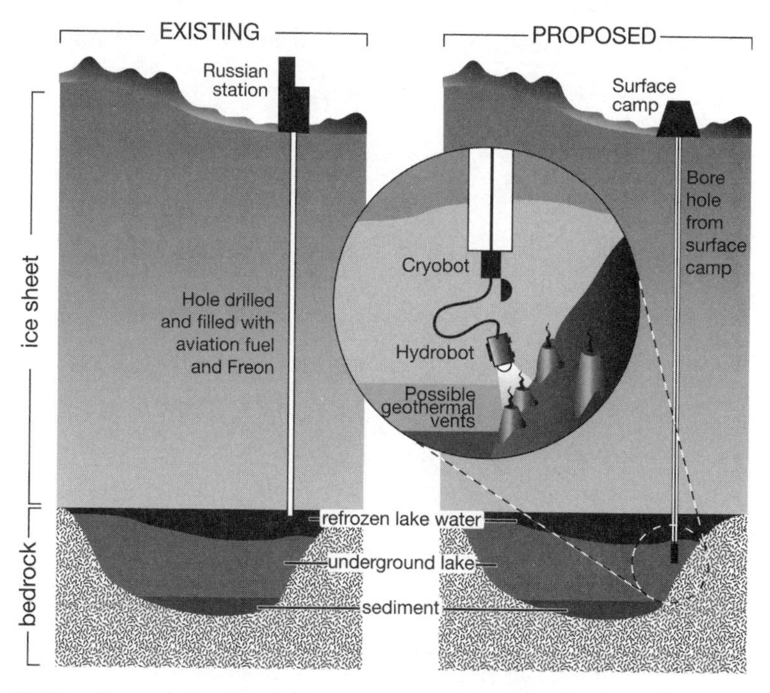

Drilling through the ice down to Lake Vostok: the present situation and the proposed solution to keeping the lake uncontaminated

life. Plans are already under way to develop a "cryobot," which will melt the ice as it bores. When it reaches the lake, the cryobot will release a small sterile "hydrobot" that will observe and collect specimens in the water, transmitting data back to the station as it moves about. These instruments are perhaps the prototypes of ones that will be placed in a spaceship and sent to Europa's surface to look for even stranger forms of life.

In less than a century we have gone from Barton and Beebe's bathysphere to hydrobots. Technology has advanced, instruments have grown more sophisticated, and we have become increasingly dependent on them. Remote sensing has replaced direct perception and yet, as the approach to Lake Vostok attests, human ingenuity, curiosity, and endurance pave the road to discovery. Barton and Beebe must be smiling in their graves.

Chapter 5

MESSAGES FROM THE SUN

I HAVE A SENTIMENTAL FONDNESS for how heat emanates from the Sun. The first real science book I ever read, at the age of thirteen, was *The Birth and Death of the Sun* by George Gamow. It tells a tale full of twists and turns, of supposedly important clues that lead down blind alleys and of seemingly irrelevant byways that point to the highway. Irreconcilable points of view are suddenly joined and data from different fields enter in unexpected ways. All this plus Gamow's own charming line drawings.

I suspect Gamow himself also had a soft spot for *The Birth and Death of the Sun.* By 1964 the field had changed so much that he wrote *A Star Called the Sun,* which begins with an inscription, "Dedicated to the memory of my book *The Birth and Death of the Sun.* Gamow will appear at many points in this chapter, not because I am trying to put him there, but because he belongs.

Born in Odessa in 1904, Gamow found the late-1920s Soviet regime increasingly oppressive. After Gamow and his wife failed in an ill-fated attempt to escape by kayak crossing the Black Sea, they managed to exit in 1933 on the pretext of a short visit to the West to attend a conference. Gamow never went back. In 1934 he moved to the United States, taking up a position at George Washington University in Washington, D.C., quickly joined there by Edward Teller, the Hungarian exile later known as the Father of the Hydrogen Bomb.

George Gamow, looking mischievous, as usual

At the Core

The fifteenth-century humanist Marsilio Ficino said in his treatise *On the Sun*, ". . . the heat which accompanies it fosters and nourishes all things and is the universal generator and mover." The Sun is the origin of almost all our heat. Energy deep within the Earth creates hydrothermal vents and volcanoes, but the Sun's energy contributes 10,000 times as much as all the geothermal energy bubbling up. Most of that energy originates in the Sun's core, but all we can see is the 5800-degree-Kelvin surface; the rest is inferred. Our knowledge of that surface temperature, deduced from the Stefan-Boltzmann Law a little more than a century ago, presents its own puzzles. For one, the surface temperature is not uniform. The Sun has dark spots—darker because they are cooler, about 4300 degrees Kelvin instead of 5800. We call them sunspots and observed them for a long time before we knew what they meant or what caused them.

Early Chinese astronomers gazing at the Sun's reflection in a lake noticed dark areas on its surface, but the systematic study of sunspots began right after the discovery of the telescope. Galileo noticed them. You might even say his troubles with the Inquisition all started with sunspots. The Jesuit Father Christopher Scheiner, one of the first sunspot observers, thought they were small planets drifting between the Sun and the Earth, not spots on the Sun. Taking issue with this identification, Galileo recognized them for what they are, tracked their motion, and observed their growth, decay, and change of shape. With his usual sureness, which opponents called arrogance if not heresy, Galileo attributed sunspot movement to the Sun's complete rotation within approximately a four-week period:

> Let them be vapors or exhalations then, or clouds, or fumes sent out from the Sun's globe or attracted there from other places; I do not decide on this—and they may be any of a thousand other things not perceived by us. . . . I shall say that the solar spots are produced and dissolved upon the surface of the Sun and are contiguous to it.

These spots were decisive for Galileo's future. His 1613 publication, *Letters on Sunspots,* contained his first statement of the principle of inertia and, more important in a political sense, his first written support of the Copernican notion of a heliocentric world. Unfortunately for Galileo, his strong claims sometimes antagonized other observers. More was at stake for the Catholic Church than the motion of sunspots. Its version of the heavens was being challenged.

We now understand that the spots are ultimately due to

the Sun's magnetic field, but it took 300 years for the connection to be made and another fifty years to understand how the twisting and flipping of that field on an eleven-year cycle can lead to eruptions on the solar surface and colder spots. The solar surface story still has its fine points that need to be worked out. As for the solar core, all models of the interior were completely wrong. The eventual solution to the puzzle of what fuels the Sun came as a surprise to everyone.

During the second half of the nineteenth century it became increasingly clear that a long period of time was necessary to account for the geological shaping of the Earth and Darwin's portrayal of the biological evolution of its inhabitants. It was hard to designate a specific number, but hundreds of millions of years seemed to be the very minimum that could account for all the changes. This meant the Sun had to be at least that old, but that was impossible if only known energy-producing mechanisms were employed; a ball of coal the size of the Sun producing energy at the observed solar rate would last thousands of years, a far cry from the required hundreds of millions.

Several of the greatest physical scientists of the time addressed the mismatch between biological and astronomical constructs of the solar lifetime. Lord Kelvin and Hermann von Helmholtz independently proposed the most promising solution, the one allowing the greatest possible solar lifetime. They argued that stored gravitational energy was slowly being converted to heat as the Sun contracted. Their answer for the solar lifetime, thirty million years, was longer than previous estimates but still not enough to explain the evolutionary phenomena Darwin had hypothesized.

The resolution started with the discovery of radioactivity, although it wasn't immediately obvious. When Pierre and Marie Curie isolated radium, the full power of radioactivity was finally revealed; a tiny source of the new material generated large amounts of heat without any apparent cooling of the radium. In 1904 Ernest Rutherford, soon to become the dominant figure in nuclear physics, was prepared to declare:

> The discovery of the radioactive elements, which in their disintegration liberate enormous amounts of energy, thus increases the possible limit of the duration of life on this planet, and allows the time claimed by the geologist and biologist for the process of evolution.

In a somewhat less formal manner, Rutherford gave an account of the octogenarian Lord Kelvin's reaction to Rutherford's views on radioactivity's reconciliation of astronomy-physics with biology-geology:

> I came into the room which was half dark, and presently spotted Lord Kelvin in the audience and realized that I was in for trouble at the last part of my speech dealing with the age of the earth, where my views conflicted with his. To my relief, Kelvin fell fast asleep, but as I came to the important point, I saw the old bird sit up, open an eye and cock a baleful glance at me! Then a sudden inspiration came, and I said Lord Kelvin had limited the age of the earth, provided no new source of energy was discovered. The prophetic utterance refers to what we are considering tonight, radium! Behold! The old boy beamed upon me.

The true source of the Sun's heat is fusion of hydrogen to helium, not radioactivity. The two are related; both are nuclear processes. Technically, Rutherford's 1904 statement is correct in that radioactivity revealed new sources, even though he did not know the specific mechanism of the release. That took another fifteen years. By then, very careful measurements had shown there was a difference between the masses of a helium nucleus and that of four hydrogen nuclei. The difference was small but far from insignificant if converted into energy. Einstein's insight of the relation between mass and energy, summarized in the famous equation $E = mc^2$, emphasized this point. In his 1920 presidential speech to the British Society for the Advancement of Science, Sir Arthur Eddington pointed out to his fellow astronomers that the combination of the hydrogen-helium mass difference and the Einstein relation meant hydrogen fusion could be the Sun's source of energy.

The Sun's core is a giant nuclear reactor in which a thousand billion pounds of hydrogen nuclei are converted every second into helium nuclei. I say nuclei rather than atoms because the ambient temperature is so high that the electrons, stripped from the atoms, constitute a kind of gas through which the nuclei move. We learned in the twentieth century that atoms are composed of nuclei surrounded by much lighter orbiting electrons. We also learned that, in addition to protons, nuclei have neutrons within them—particles tightly bound to protons, constituting a kind of nuclear ballast, with mass but no electric charge.

Each pound of hydrogen produces as much energy in fusing as the burning of twenty million pounds of coal, employing the same process as the one that fuels the hydrogen bomb. In order to perform the hydrogen-to-helium conversion, the core's central temperature needs to be over fifteen million degrees Kelvin, a number arrived at by our understanding of solar dynamics and the nuclear physics of fusion mechanisms. Reassuringly, the Sun has had enough hydrogen to fuel this process for billions of years, long enough for simple single-celled organisms to have evolved into primates, long enough for the building-up of complex civilizations.

By the late 1930s, the basic principle of how the solar core produces energy was understood. Still missing, however, was a detailed description of the chain of nuclear reactions that included temperatures and pressures. In Washington, our friend George Gamow decided the problem was ripe for a solution. Together with Teller and Merle Tuve, he had convened on a yearly basis meetings to discuss intriguing topics in physics. The 1938 choice was astrophysics and, in particular, the mechanisms of solar-core workings.

The Washington group called up their colleague Hans Bethe, the great master of nuclear physics and the recent author of the three articles on the subject known as "Bethe's Bible." As Bethe remembers, "I was totally uninterested in astrophysics, so I said I wouldn't come. Edward Teller persuaded me nevertheless to come and it turned out to be probably the most important conference that I attended in my life." Over a six-month period after the meeting, Bethe worked out the details of the nuclear reactions that fuel the stars and earned a Nobel Prize in the bargain. Which particular nuclear reactions catalyze the hydrogen-to-helium fusion and at what rate they proceed depends on the core temperature of the Sun.

My fascination with the Sun's workings, begun at age thirteen, has not waned. One of the most remarkable observations of a solar property ever made was not in Gamow's first book about the Sun and it wasn't in his second one, either. It's too recent: a direct temperature measurement of the Sun's very core.

Cosmic Gall

Since most solar energy is produced in the Sun's core and then diffuses out over a period of millions of years, it's clear that the core holds the key to what powers the Sun. The Earth's core temperature is not that well known: estimates range between 5000 and 6000 degrees Kelvin, a 20 percent difference. Somewhat miraculously, we know the Sun's core temperature far more precisely than that. Furthermore, we have a telescope measurement that provides us with solid data confirming our estimates.

It's not an optical telescope because you can't see the center of the Sun, at least not "see" in the usual sense of the word. We are very fortunate, however, in having a messenger from the Sun's core, a particle produced in its very heart, that tells us the core temperature. This speedy messenger reaches the solar surface a little over two seconds after its creation, and reaches the Earth 8.3 minutes later. It's called a neutrino.

You will never see, feel, smell, or touch a neutrino. Thousands of them, once produced in the Sun or in faraway stars or perhaps even in the first seconds after the Big Bang, go through your hands every second. You will never be aware of their passage, they will never change your life in the slightest way, and you will not live a second longer or less because of their passage through you. Had it not been for some mysterious results in 1930s nuclear physics experiments, nobody would have ever thought of them, and yet neutrinos turn out to be key players in describing the evolution of the universe. Science leads you to think about some strange things, and few are stranger than neutrinos. I and a few thousand other people have spent much of our professional careers trying to understand their properties. We consider ourselves truly fortunate to be allowed and even encouraged to think about and measure the properties of one of the universe's most exotic constituents.

Forty years ago, the then very young John Updike described their behavior in verse:

Cosmic Gall

Neutrinos, they are very small.
They have no charge and they have no mass
And do not interact at all.
The Earth is just a silly ball
To them, through which they simply pass,
Like dustmaids down a drafty hall
Or photons through a sheet of glass.
They snub the most exquisite gas,
Ignore the most substantial wall,
Cold shoulder steel and sounding brass,
Insult the Stallion in his stall,
And, scorning barriers of class,
Infiltrate you and me! Like tall
And painless guillotines, they fall
Down through our heads into the grass.
At night, they enter at Nepal
And pierce the lover and his lass
From Underneath the Bed—you call
It wonderful; I call it crass.

When I was starting my career, the common assumption was that the neutrino had no mass, as Updike says in his second line. We now think it has a very small mass, but we don't know how small. That mass, perplexingly minute, is less than one ten-thousandth as big as the electron's, the smallest directly measured particle mass. The neutrino certainly has no electric charge, but it isn't quite right to say it does "not interact at all."

Rarely—even very, very rarely—a neutrino meeting a neutron will metamorphose into an electron, in the process converting the neutron into a proton. Out of many, many neutrinos passing by many, many neutrons, one might make the change. The ways and means by which this happens were suggested in 1934 by Enrico Fermi, who coined the name *neutrino,* little neutral one, to distinguish the particle from the very massive, at least by comparison, *neutrone,* or big neutral one.

Fermi's paper on the subject was rejected by *Nature* because "it contained speculations too remote from reality to be of interest to the reader." *Nature* was wrong; the paper was one of the greatest and the rejection is now remembered as a case of poor editorial judgment. Though still remote from reality, neutrinos are now observed routinely in high-energy accelerator experiments and emerging from nuclear reactors. Recently, they have even been detected leaving the Sun's core and used to measure temperature.

This is a story of high-tech modern science. Many, perhaps most of us, occasionally look back with nostalgia to the days when machines were simple and all measurements were direct, but forefront science doesn't advance with rulers, clocks, and thermometers. The STM (scanning tunneling microscope) and the femtosecond laser are modern tools for measuring length and time. Temperature also has a variety of new ways of being recorded. None is more exotic, none more removed from our everyday sensory experience, than the mysterious neutrino, but that solar-core temperature is ultimately the key to the existence of day-to day life on Earth.

The story starts about the same time John Updike was writing his poem. Two physicists, John Bahcall and Ray Davis, published back-to-back papers suggesting a way to detect high-energy neutrinos coming from the center of the Sun. The particular neutrinos Bahcall and Davis were seeking are emitted in a rare neutrino-producing high-temperature nuclear decay. The only place in the Sun where all the necessary conditions for the production of those neutrinos can be met is in the Sun's very center and then only if the temperature is high enough. Observing those neutrinos measures the Sun's core temperature.

Surprisingly, the best place on the Earth to detect solar neutrinos—in fact, the only place so far—is deep underground. The reason for looking there instead of on the Earth's surface is that neutrinos easily traverse the mile of rock that stops other messengers from outer space. On the surface, it is impossible to distinguish a rarely occurring neutrino signal from other more frequent ones.

Bahcall, the theorist, estimated the expected solar neutrino flux from the nuclear decays, and Davis, the experimentalist, described the tool needed to find them. Davis chose an Olympic-sized pool full of a chlorine-rich substance for his measurements. The energetic solar-core neutrinos very occasionally change a chlorine nucleus into an argon nucleus. In the process, a neutron inside the chlorine nucleus turns into a proton. To do the experiment, Davis needed a lot of chlorine in a liquid form: perchloroethylene—cleaning fluid—cheap and readily available, was the obvious liquid. As he prepared the apparatus, commercial distributors, judging the size of his order, thought he was running a chain of dry cleaning stores. Unfortunately, those neutrino reactions are so scarce that, even with an Olympic-sized pool full of chlorine, Davis expected to see only a single argon atom a day. By comparison, looking for a needle in a haystack is child's play. But Davis, then at Brookhaven National Laboratory and now a colleague of mine at the University of Pennsylvania, thought he could find that one atom.

The chlorine tank had to be placed deep underground to ensure that the argon was produced by solar neutrinos, not by another source; the deeper the placement, the safer the measurement. This meant looking for a gold mine, because the world's deepest mines are gold mines. With the aid of government funding, Davis managed to assemble and fill the tank in a chamber provided by South Dakota's Homestake Gold Mine. Trained mining personnel supplied support; all the equipment needed to be lowered on the same elevator the miners had used. Physicists and miners wore the same helmets, one group looking for scientific gold with their neutrino telescope, the other looking for a more traditional gold.

As the experiment continued over a twenty-year period, Davis and collaborators found one argon atom every three days instead of one every day. The expected neutrino-counting rate depends on the twenty-fifth power of the Sun's core temperature. If the Sun's central temperature is a little higher than Bahcall's estimate, you will see more than one argon atom a day. If it is only 10 percent lower than the estimate, you expect one argon atom every three days. Most neutrino experts, myself included, assumed there was either a flaw in the experiment or a small error in the solar model's estimate of the Sun's core temperature. We were wrong.

Newer detectors, operating totally differently, are even capable of seeing the direction neutrinos are coming from. Their results agree with Davis's experiment; Bahcall's calculations are also now known to be correct. Apparently, something funny is happening to neutrinos on their way to us from the solar core. The preferred explanation, confirmed by very recent measurements, is that some neutrinos produced in the core, perhaps two-thirds of them, change identity during the two seconds it takes them to reach the solar surface. They are produced as expected, but Davis's apparatus is sensitive only to the unchanged ones.

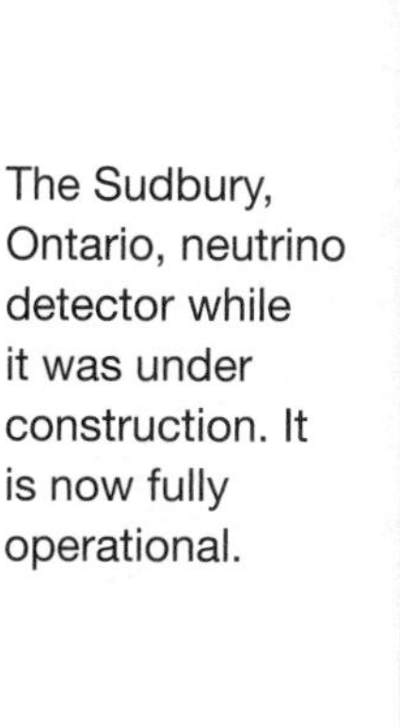

The Sudbury, Ontario, neutrino detector while it was under construction. It is now fully operational.

All these puzzles are now being studied in large, deep, underground laboratories in the Japanese Alps, under Gran Sasso Mountain near Rome, in the Urals at Baksan, in Sudbury, Canada, and in a swimming pool filled with cleaning fluid in South Dakota. The concept of a neutrino telescope never ceases to amaze me. The very weakness of the neutrinos' interactions means they stream straight out of the Sun's interior and through the Earth. Only one of the neutrinos leaving the core in a three-day period stops in that tank. But those neutrinos tell us the Sun's core temperature is 15.7 million degrees Kelvin. The accuracy of that measurement is better than 1 percent!

A Thermal Aside: Gamow, Rutherford, and Nuclear Barriers

Why does neutrino emission from the Sun's core depend so sensitively on temperature? The answer involves quantum mechanics, the beginnings of experimental nuclear physics, and two of my favorite physicists.

Ernest Rutherford, whose parents were among New Zealand's original settlers, appeared already in this chapter proclaiming radioactivity as the source of the Sun's heat. In the late 1890s, Rutherford discovered that helium nuclei, then known as alpha rays or alpha particles, come out of radioactive sources. Fifteen years later, Rutherford began bombarding a thin gold foil with alpha rays in order to study the atom's structure. The then prevailing theory was formulated by J. J. Thomson, head of the Cavendish Laboratory at Cambridge, a Nobel Prize winner, and Rutherford's mentor and former teacher. Thomson's "raisin pudding model" of the electrically neutral atom has a uniformly positive background with small negative electrons embedded in it. According to Thomson, alpha rays would pass right through the gold foil, undeflected.

Contrary to his mentor's expectations, Rutherford found alpha rays occasionally scattering through wide angles and sometimes even bouncing right back. As Rutherford put it, "It was as though you had fired a fifteen-inch shell at a piece of tissue paper and it had bounced back and hit you." There was one possible explanation: if all the atom's positive charge is concentrated in a tiny positive core—Rutherford called it "atomic nucleus"—the positively charged alpha rays will be essentially undeflected unless they come close to the core. If they do, you expect a big electrical repulsion and big scattering. How tiny is the nucleus? Typically, its diameter is a few, perhaps ten, millionths of the atom's size.

By 1928 Rutherford was a legend, the winner of the Nobel Prize in chemistry, the world's top experimental physicist, the discoverer of the atomic nucleus, and Thomson's successor as director of Cambridge's Cavendish Lab, the premier nuclear physics lab in the world. Still very active in research, Rutherford was puzzled by the details of nuclear reactions, the processes that heat the Sun's core. In studying radioactivity, Rutherford observed alpha particles, helium nuclei, leaking out from inside larger nuclei. He asked himself: Are they inside for a while and then somehow outside, or are they created while something else is coming out? How do we explain this leakage?

A great experimentalist, Rutherford was often suspicious of the mathematical wizardry of the young cadre of quantum mechanics practitioners. In particular, they hadn't answered his puzzle. Groping for an explanation, he wrote a paper laying out the parameters of the nuclear decay problem. Gamow, then in Copenhagen, read the paper and thought he had the answer. He talked to Bohr, who sent him on to Cambridge. So it came that one day in early 1929 Rutherford saw appearing on his doorstep a very tall twenty-four-year-old Russian, speaking only broken English. Gamow promptly explained to Rutherford how to solve the decay problem according to the new rules of quantum mechanics. His "Gamow Factor," is the quantum-mechanical probability of a particle penetrating through a barrier.

The helium nucleus is kept inside the larger and heavier nucleus by an effective nuclear force, creating a kind of barrier. But, as Gamow put it, "In quantum mechanics there are no impenetrable barriers." Moreover, the time dependence for leaking through the barrier, for the helium nucleus to appear on the outside, comes out just right when calculated using quantum mechanics.

Ernest Rutherford (right) with his assistant, J. A. Ratcliffe, in their laboratory, standing next to the apparatus for detecting alpha particles

There are really two barriers, and it's important to distinguish them. The nuclear force is much stronger than electrical forces but operates only over very short distances. This force holds protons and helium nuclei packed together inside the larger nucleus. Unconstrained, they would fly apart because, being positively charged, they repel each other electrically.

That repulsion is overcome by the nuclear force, leading, of course, to a barrier against exit from the nucleus. But, as Gamow said, there are no impenetrable barriers. Nuclei do decay.

The first kind of barrier, strong force, succeeds most of the time in keeping protons or helium nuclei from getting *out* of a larger nucleus. The second kind, electrical force, prevents outside protons or helium nuclei from getting *into* a nucleus, again most of the time. An outside proton or a helium nucleus approaching a larger nucleus will feel the long-range repulsive electrical force of the nucleus's constituents well before it experiences any attraction due to the short-range nuclear force, and will normally veer away. The higher the ambient temperature, the more energetic the outside projectile, and therefore the likelier it is to continue on its path toward the nucleus. At first the growing electrical barrier appears insurmountable.

Gamow, however, emphasized to Rutherford that this barrier can be penetrated. If the nucleus is unstable after the entry, the process will stimulate its breakup. The temperature sensitivity of barrier penetration is ultimately the cause of the temperature sensitivity of neutrino emission from the Sun's core. Performing laboratory experiments to explore barrier penetration helped persuade Rutherford to launch the construction of accelerators for the disintegration of the nucleus. That's another story, one with its own twists and turns.

The term for a nucleus splitting up is "fission," and that for nuclei coming together is "fusion." Both occur, and temperature plays a role in both, particularly in fusion. At the high-temperature and high-pressure conditions that prevail in the Sun's core, protons that repel each other electrically are squeezed together and fuse into helium nuclei, liberating in the process the vast amounts of energy that power our Sun.

A Star Is Born

The temperature at the center of the Earth is thousands of degrees. In the center of a small to midsized star like the Sun, it is millions of degrees. A larger star, say twenty-five times as massive as the Sun, reaches a central temperature of billions of degrees. What happens then? What kinds of neutrinos, if any, does the big star emit? How do we measure its core temperature? Let's look at some historical records that bear on the birth and death of stars.

According to Aristotle's tenets, a star could neither die nor be born. This view still held in sixteenth-century Europe. Tycho Brahe, one of the last great pretelescope astronomers, took a walk after dinner on the evening of November 11, 1572. Looking at his familiar night sky, he noticed that, contrary to Aristotle's dictum, a new star had appeared. There was no possibility of a mistake. As he later told the story,

> Amazed and as if astonished and stupefied, I stood still, gazing for a certain length of time with my eyes fixed intently upon it and noticing that same star placed close to the stars which antiquity attributed to Cassiopeia. When I had satisfied myself that no star of that kind had ever shone forth before, I was led into such perplexity by the unbelievability of the thing that I began to doubt the faith of my own eyes.

Tycho had seen what we now call a *supernova,* something new appearing in the skies. Though Tycho didn't know it, he wasn't the first to see one. Chinese and Japanese astronomers recorded the appearance of a bright new star in the Crab Nebula on July 4, 1054. As Yang Wei T'e, imperial astronomer to the Sung Dynasty, recorded:

> I make my kowtow. I observed the phenomenon of a guest star. Its color was slightly iridescent. Following an order of the Emperor, I respectfully make the prediction that the star does not disturb Aldebaran. This indicates that the Emperor will gain great power.

An Arab text records an *athari kawkab* (spectacular star) in the same summer, and a petroglyph in Navajo Canyon, Arizona, may also be a register of the event. Other supernovas were surely witnessed by our ancestors. It has even been suggested that the bright star guiding the Three Kings toward the infant Jesus was one. The first visible supernova explosion after Tycho's observation came a mere thirty-two years later; Kepler not only saw the star but followed it for a year before it began to fade. Galileo gave lectures about it in Padua.

In many ways, a star's life is very simple, certainly simpler than its death. Gravity's attractive force pulls the star inward, creating mounting pressure, many billions of times greater near the center than at the surface. To counterbalance the inward pressure, heat generated in the core pushes back. The bigger the star, the hotter the core needs to be to prevent collapse. Balance between the two, inward gravity and outward heat, holds until the end of a star's life.

Stars like the Sun, or a few times bigger, live by fusing hydrogen; later in the life cycle, when pressures and temperatures are high enough, helium nuclei fuse into carbon or oxygen nuclei. Four hydrogen nuclei take up more room than a helium nucleus; the change leads to compression, a squeezing together that heats the core. This raises the temperature and increases the burning rate and the luminosity. Scientists believe the Sun has grown brighter for 4.5 billion years, increasing its energy output over this span of time by almost 30 percent.

This increase of solar energy output causes some problems for our view of the Earth's past. A cooler Sun in the past means a cooler Earth. If no other factors intervened, the Earth's oceans would have remained frozen until about 1.5 billion years ago, but that can't be true. Liquid water is known to have been present at least two billion years earlier. The most likely solution to the dilemma is carbon dioxide levels a thousand times greater than what we have now. A feedback loop probably brought those levels under control once the Earth's temperature rose. This mechanism kept the oceans from freezing when the Sun's luminosity was low and later restored the carbon dioxide levels to values more like the present ones.

Rising solar luminosity also leads to some forecasts of what we might expect the Earth's surface to be like in the future. It's not a pretty picture. By about a billion years from now, temperatures will rise to the boiling point of water; the only potential survivors will be hyperthermophilic bacteria. Shortly after that, the Earth will go the way of Venus, with soaring temperatures and a toxic atmosphere. At that point, all forms of life on Earth will cease. The five-billion-year record of living organisms will be expunged. The Earth itself will not end; it will go on, lifeless, for a few billion years after that.

There is a small caveat, recently proposed by Donald Korycansky of the University of California at Santa Cruz. If an asteroid about fifty miles across and weighing a little more than a million billion tons could be harnessed, it might be able to nudge the Earth outward in the solar system once every 6000 years, thereby correcting for increasing solar luminosity. The asteroid needs to impart some energy to the Earth, then shoot out past Jupiter to regain energy. After a rocket-induced change of course, it returns to the vicinity of Earth, giving our planet another boost, and then restarts the cycle. In a few billion years, the Earth will have moved close to the present orbit of Mars. Sounds unlikely, but interesting to contemplate.

As for the Sun's projected final stages, once the hydrogen in the core is consumed and a maximum temperature reached, the outer shells of hydrogen will begin to burn. The Sun will swell up, forming a cloud that cools, turning from yellowish to reddish, from 5800 degrees to 3500 degrees Kelvin at the surface. It will now be a red giant, a hot, diffuse gaseous cloud stretching out to the present orbit of Mars. When essentially all the fuel is exhausted, the Sun will begin to cool and contract. By then the Earth will have disappeared, engulfed and evaporated by the hot cloud, perhaps leaving behind a trace of the rocky grain that grew into Earth almost ten billion years earlier. The shrinking Sun, once an ordinary star and then a red giant, will have a renewed life as a shrunken white dwarf star before finally fading into oblivion.

Bigger stars, with higher core pressures and temperatures, have a lengthier chain of nuclear reactions than the Sun and a different death. The death I am going to describe, that of stars ten to forty times as massive as our Sun, has already taken place a hundred million times in our own galaxy. The typical central "compress and heat" scenario brings the star's core to higher and higher temperatures for shorter and shorter times. At each stage a heavier kind of nucleus dominates the core contents. At forty million degrees it's helium, then carbon, neon, oxygen, and silicon. By then the central temperature has reached billions of degrees. The final phase, the silicon-to-iron conversion, lasts only a day.

At each step, energy is released, sustaining the star for a while. But the series ends. Iron nuclei need energy to fuse, but none is available. With no next step in the nuclear chain, no outward pressure to counteract the squeeze of gravity, the star begins a rapid collapse. In a few seconds, the core, originally larger than the whole Earth, implodes to a kernel no bigger than a midsized city. Meanwhile, the nuclei produced earlier, now arranged in onionskin-like layers about that central core, feel the effect of the collapse. Left unsupported, they come crashing in and then bounce back, after hitting the wall of the newly formed core. They are distributed into space, leaving behind the remnant of a once great star.

Iron, magnesium, sulfur, carbon, oxygen, neon, and all the other elements in the universe, except hydrogen, helium, and small amounts of lithium and beryllium, owe their existence to these explosions. These elements are forged in the great stellar furnaces and dispersed by the stars' sudden deaths. Our own Earth is made up almost entirely out of material produced inside these large stars and then scattered into space as those stars came to a fiery end.

The explosion associated with the death of these stars is the largest in our cosmos. For months the explosion's glow is brighter than the billions of other stars in its galaxy combined. Bright as it is, we "see" only 1 percent of the energy released in the explosion. During the core's collapse, electrons and protons are squeezed together, each pair forming a neutron and a neutrino. The neutrons push together, forming a fluid so dense that a teaspoonful weighs a million tons. The neutrinos now depart from the core. Not right away; even elusive neutrinos are trapped for a while. Capable of effortlessly gliding through what Updike called the "silly ball" of Earth, they too feel the pressure of the ultradense surroundings. However, neutrinos do escape, carrying off 99 percent of the energy released in the explosion. High density means they need ten seconds to leave the core, a distance they would normally travel in a microsecond. During those seconds they also reach thermal equilibrium with their surroundings. Once departed, the neutrinos bear with them the memory of the multibillion-degree temperature within the collapsing core.

In 1940 George Gamow and the Brazilian physicist Mario Schoenberg wrote the first paper to consider a burst of neutrinos as a cooling mechanism for a star's shrinking core. With his unfailing humor, Gamow labeled the process URCA, after the Rio de Janeiro Casino de Urca, where money flowed quickly out of gamblers' pockets. Concerned that the journal to which the paper was submitted, *Physical Review,* might question the origin of the acronym, Gamow prepared a backup story where URCA stood for "unrecordable cooling agent," unrecordable, of course, because one never expected to see the neutrinos. Fortunately, there were no questions and the name URCA stands.

Given the number of exploding stars of this type, a hundred million or more in a five-billion-year history, we can expect to see one with the naked eye every few hundred years, just as Yang Wei T'e, Tycho Brahe, Kepler, and Galileo had. Observing the exploding star's core neutrinos and determining its core temperature is a good deal harder.

Black Holes and Little Green Men

Until 1987, no sudden stellar appearance had been seen in either our galaxy or the nearby Magellanic cloud since Galileo and Kepler's 1609 viewings. But on the evening of February 23, 1987, a Canadian astronomer named Ian Shelton was studying stars in the Large Magellanic Cloud. He saw a bright spot on one of his photographic plates that he hadn't noticed before. He thought it was a mistake, a defect in the plate. In any case, the night sky from his observatory in the high Andes was exceptionally clear, so Shelton thought it was worth checking his plate. He stepped outside. This was no plate defect: a new bright star had appeared.

A supernova was shining, a new nearby light was in the sky—an event like the one Tycho Brahe witnessed 415 years earlier. Shelton, on his perch in the Andes, reported it. So did Albert Jones, a New Zealand amateur astronomer who spotted the glow an hour before Shelton. Word quickly got out and astronomers around the world scanned their plates, trying to pin down the star's first appearance. Older plates showed a blue supergiant, Sanduleak-69202, at the same spot, its death the source of the bright new light. The first photographic film record of the supernova is dated February 23, 1987, at 10:38 Universal Time (also known as Greenwich Mean Time).

Jones and Shelton witnessed the supernova explosion on February 23, 1987, but the event actually occurred much earlier. The star is so far away that its light takes 170,000 years of travel to arrive on Earth. To appreciate how far away that is, remember that sunlight reaches us in only eight minutes. The event's brightness reminds us how powerful the explosion was. Remember, however, that 99 percent of the energy released in the explosion is carried off by neutrinos slicing through the star's outer layers. Since neutrinos are presumed to travel at almost exactly the speed of light (perhaps a tiny bit slower, but that's a technical point), they also take 170,000 years to reach us from that star.

There is one very important difference between the light signal and the neutrino one. Light diffuses from the cloud surrounding the explosion over the course of months, perhaps years. The first light only appears hours after the collapse. Neutrinos come out in a single ten-second burst. The building and maintaining of the apparatus needed to detect a ten-second burst of neutrinos from an object that far away is no easy matter. It almost certainly would not have been built if that's all it could do, because such supernova events are only expected every few hundred years. Fortunately, very fortunately, detectors in some ways like Ray Davis's were set up in the early 1980s at the Morton salt mine in Ohio, under Mount Kamioka in Japan, and at Baksan in the Russian Ural Mountains. Their primary aim was to look for the very rare conjectured decay of the proton. With foresight, their instrumentation was also capable of detecting the short burst of neutrinos from a relatively nearby supernova explosion and timing their arrival. A group from my university had been instrumental in setting up the Mount Kamioka apparatus, so I was aware of its capabilities, but didn't expect a supernova.

The apparatus is designed to record events electronically even if nobody is present. In February 1987, after the optical sighting, the personnel at each detector went to check recorded events, hoping to see between five and ten neutrino-initiated signals. To set the scale, roughly a thousand billion billion billion billion billion billion neutrinos left the supernova, distributed evenly over the sky. A thousand billion billion billion of those reached the Earth: the prediction was that a few would make their presence known in each detector. The neutrinos were there, right on schedule; a ten-second burst appeared in each detector on February 23, 1987, at 7:35:40 Universal Time, a little over three hours before the first light appeared on photographic plates. Kamioka recorded eleven events, Morton eight, and Baksan five. I was in England at the time. I received an excited phone call from one of my colleagues (it would now be an e-mail) telling me they found the neutrino signal in their Kamioka data. I remember Nobel Prize winner Carlo Rubbia saying, "The field of extra-solar neutrino astronomy went from science fiction to science fact." It all happened in ten seconds.

It isn't just the number of neutrinos and their timing that agrees with predictions. When the neutrinos escaped from the core, they were in thermal equilibrium. The Baksan, Kamioka, and Morton groups studied their data, reconstructing neutrino energies, presumably unchanged during the 170,000-year transit. The analysis showed that the neutrinos reflect a one-hundred-billion-degree thermal distribution, the core temperature when the neutrinos left. The agreement of theory and evidence is almost too good to be true.

This still leaves us with some puzzles. The supernova blast releases a tremendous amount of energy, and yet the nuclear fuel in the star is used up. So what powers the explosion? Ironically, Lord Kelvin and Hermann von Helmholtz independently provided the resolution to this question while suggesting a mechanism to explain the Sun's lifetime, a hundred years before anybody considered rapidly collapsing stars. While gravitational contraction, the Kelvin-Helmholtz mechanism, is an insignificant contributor to our Sun's energy output, the change in gravitational energy during the shrinkage of a large star's core down to a ten-mile radius is truly prodigious. By conservation of energy, the decreasing gravitational energy has to reappear in another form. Part of it is converted into radiation, but most of it surfaces as motion energy of the neutrinos. In ten seconds, they carry away as much energy as the Sun releases in ten billion years.

The core temperature at the time of collapse is a hundred billion degrees, but there seems to be no reason why this should be the ultimate limit. Why not a trillion degrees, a hundred trillion? After all, the Kelvin-Helmholtz heating scheme is just conservation of energy. As surely as a ball rolling down a hill has its potential energy converted to kinetic energy, a neutron from the outside of a star gains energy as it falls toward the center. A star fifty times as massive as the Sun might yield an even hotter core before exploding. It doesn't. The endpoint of that fifty-solar-mass star's life is a black hole, the perfect absorber, emitter of nothing. Everything goes in; nothing comes out.

When enough mass is shrunk into a small enough region, nothing on the surface escapes the growing gravitational field. By small enough I mean shrinking the Sun down to a sphere a mile across, or the Earth into one an inch across. These are obviously almost unimaginably high densities, but there is nothing in the laws of physics that excludes their possibility. You might think a messenger launched with high enough speed or light itself, carrying no mass, could escape. Those loopholes have been closed by Einstein's insights into the behavior of matter and light. In formulating his Special Theory of Relativity, Einstein realized no signal could travel faster than the speed of light. A massive object would have to travel faster than light to escape, but that's impossible. As for light, by relating mass to energy, Einstein showed that gravity's primary sensitivity is to energy, not simply mass. Light carries energy so it also can't escape. A black hole is a black hole. Nothing comes out of it.

Serious scientific discussions about black holes surfaced in the 1930s, but the possibility of ever seeing any evidence of one seemed too remote. The step before black hole formation, the creation of a ball of neutrons in the core of a massive star, though perhaps not quite as bizarre, seemed almost as difficult. Radiation from a star only a few miles in diameter was simply not in the astronomy lexicon until the 1960s.

In late 1967 a young Cambridge University graduate student named Jocelyn Bell was studying radiation reaching an array of radio telescopes covering more than four acres of English countryside. She noticed that one point in the sky was emitting an extraordinarily regular signal, arriving every

1.3373001 seconds. She and her thesis adviser, Anthony Hewish, held off publication for a while, lest the public jump to the conclusion that aliens were contacting us. At first they labeled their find with the whimsical code name of L.G.M. (Little Green Men). After finding a second and a third Little Green Man with different periods, the astronomers came to believe the message was from a star, not from an alien reaching out.

It was clear that whatever kind of star it was couldn't be bigger than a few hundred miles in diameter. The transmitter completed a full revolution in little over a second; the surface of anything much larger, spinning that quickly, will break apart. A white dwarf, whose typical diameter is 5000 miles, was quickly ruled out. Its edge would have to be moving at almost the speed of light in order to complete a full 360-degree turn in a second. The news was particularly unsettling because most astronomers at the time thought that stars in their death throes shed enough matter for reincarnation as white dwarfs.

The choice was narrowed to neutron stars or black holes, but nobody suspected that either one of them could radiate huge amounts of energy on a regular basis. The core left over after a supernova explosion is small, cold, and spinning very rapidly. Just as a skater's twirling accelerates when the skater pulls his or her arms in, a star's rotating speeds up as it contracts. By the time it is ten miles across, it can easily be performing a full revolution in a second, but how does a neutron star radiate? This is where another surprise came in, a detail glossed over in the first studies of neutron stars.

The Earth has a magnetic field; the Sun has a much stronger one. If you compress a star, you also compress and amplify the magnetic field enormously in the transition from a large star to a neutron star. A rapidly rotating neutron star's magnetic field also points at an angle with respect to the axis of rotation, just as the Earth's North Pole and the magnetic pole are not aligned. When the neutron star's field points at us, perhaps once a second, it acts like the beacon of a lighthouse, shining its radiation toward us. To remind us of this strange capacity and of our own detection of pulses, rapidly rotating neutron stars are now often simply called pulsars.

We used to think that pulsars had the universe's largest magnetic fields. In 1998 astronomers discovered "magnetars," stars with even bigger fields. Maybe their magnetic fields are amplified past the level of pulsars by a winding up similar to the twisting that creates sunspots, or maybe it's compression like the fields in pulsars, or maybe it's some other mechanism. Perhaps the clues are there and we haven't recognized them yet. We have been studying sunspots for three hundred years, pulsars for thirty, and magnetars for only three. The pace of scientific research is speeding up.

Once pulsars were seen, black holes came under increased scrutiny. Does it make any sense to discuss their temperature? At first, the answer is no. A star's surface temperature is determined by measuring radiation from the surface, but black holes don't radiate. Their gravitational field is so strong nothing escapes. That's the way things stood until 1975, when Stephen Hawking found a glitch in the argument, a quantum mechanical subtlety that allows black holes to radiate and even have a temperature—the Hawking temperature.

Hawking went a step further, tying the black hole's entropy, a measure of the amount of internal information, to his notion of temperature. For ordinary black holes, the kind of radiation Hawking discusses is unmeasurably small, but the link between black holes and information has turned out to be an important testing ground for superstring theory notions about how to join together quantum mechanics and general relativity. The seemingly immeasurable temperature of a black hole has already led to new mysteries. It may also provide the clues to their answers.

Understanding the temperature of stellar interiors, of collapsing stars, and perhaps even of black holes has emboldened us to look back in time to the era before stars formed, to study the temperature of the early universe and see how it shaped everything that followed.

The Founding Elements: Hydrogen and Helium

The early universe's chronicle can be narrated by growth in size, by time elapsed, or by decrease in temperature. It's the ruler, the clock, and the thermometer all over again. In our universe's case, these three tell the same story. Space stretches as time passes and temperature drops as space stretches. Only one of the three needs to be specified, and cosmologists prefer temperature.

Steven Weinberg, a physics Nobel Prize winner and one of the past half-century's most influential theoretical physicists, has a long-time interest in cosmology. In 1976 he wrote a brilliant popular book, chronicling the universe's temperature in the first three minutes and describing the work of many talented and forward-looking scientists. His narrative in *The First Three Minutes* of the temperature chronicle starts at one hundred billion degrees:

> The universe is simpler and easier to describe than it ever will be again. It is filled with an undifferentiated soup of matter and radiation, each particle of which collides very rapidly with the other particles. Thus, despite its rapid expansion, the universe is in a state of nearly perfect thermal equilibrium.

At this time, one hundredth of a second after the Big Bang, the number of neutrons and protons in the universe was equal. Were it not for the small mass difference between neutrons and protons, about one part in two thousand, and for the decay of free neutrons, the number of neutrons and protons would still be equal. As far as we know, protons live forever; neutrons do not. According to Fermi's 1934 neutrino theory, a neutron decays into a proton, an electron, and a neutrino unless the neutron is safely locked away in a nucleus. Once inside a nucleus, it acquires permanence, except in the rare instance of radioactive nuclei, an unessential complication in this case.

Essentially all those early neutrons combined with protons, two of each, to form helium nuclei, very stable forms of matter. However, the neutrons and protons couldn't bind until the universe was cool enough for incipient nuclei not to be blown apart by radiation. That point was reached some four minutes after the universe's birth, when the temperature had dropped to a billion degrees Kelvin. The ratio of primordial hydrogen to helium nuclei was then set once and for all at roughly ten to one.

Hydrogen and helium atoms formed hundreds of thousands of years later, when the ambient temperature dropped to 3000 degrees Kelvin. The reason nuclei form at billions of degrees and atoms at thousands is quite simple: radiation in thermal equilibrium has both its energy and its density fixed by temperature. If the early mix is too hot, radiation smashes apart any emerging structures. Nuclei, with their stronger cohesive forces, already resisted the onslaught at a billion degrees. Atoms, held together only by the weaker electron-nucleus electrical attraction, appeared much later. The moment when atoms formed, 300,000 or so years after the Big Bang, was the last time the universe's visible matter was in thermal equilibrium, the last time the universe could be described in its entirety by a single temperature.

A billion years later, small fluctuations in the density of matter began to grow into larger agglomerates. Clouds of hydrogen and helium formed and stars, also made of hydrogen and helium, appeared in their centers. The larger stars created heavier nuclei in their cores and then exploded, littering the interstellar medium with material that eventually made its way into newer stars.

Nevertheless, the universe remains mainly hydrogen and helium. Simply check the Sun's chemical composition. Even though our Sun is a relative youth, five billion years old in the universe's total of fifteen billion, it is 91 percent hydrogen and 9 percent helium, a ten-to-one ratio. All other elements contribute less than 1 percent.

Don't look for confirmation of the ten-to-one ratio on Earth, an anomalous rocky mound forged from supernova scraps. Tycho Brahe and Ian Shelton, looking at a new supernova explosion in the sky, were standing on the cinders of an earlier one. As often happens in cosmology, certain old phrases acquire an eerie resonance, blasphemous if heard the wrong way, but here intended as humbling. "Ashes to ashes, dust to dust" describes the human passage through life, and also the Earth's journey in the life of the universe. Our Earth was born from the ashes of a supernova explosion; Earth's very core is a mixture of iron and iron-nickel alloys, elements whose nuclei once lay in the cores of those massive stars. One day, when our Sun has turned into a growing red giant, our Earth will once again turn into ashes.

That's roughly the temperature chronicle of our universe, from hundreds of billions of degrees to thousands of degrees, but where's the proof? One piece of evidence comes from the agreement of the universe's hydrogen-to-helium ratio with the cooling scenario's prediction. Another more direct proof appeared in 1964 and changed the field of cosmology forever.

Three-Degree Photons, Two-Degree Neutrinos

In the early 1950s, Gamow and two younger collaborators, Ralph Alpher and Robert Herman, considered a model describing a very hot early universe filled with neutrons and radiation in thermal equilibrium. The neutrons begin decaying right away into protons, electrons, and neutrinos. As the environment cools, protons and the neutrons that haven't decayed join together to form nuclei and eventually the electrons attach to nuclei to make atoms.

We now know their scenario is incorrect. The universe begins with equal numbers of protons and neutrons, not just neutrons; electrons and neutrinos are there from the beginning, as is the radiation. Nonetheless, Alpher, Gamow, and Herman's work was the first serious attempt to discuss the observable consequences of an early explosion of space leaving in its wake an expanding cooling universe. The name "Big Bang" was coined by astronomer Fred Hoyle to make fun of these early efforts. Ironically, the name stuck.

The trio even asked if any radiation from the early universe might still be visible. The Big Bang's remnants weren't looked for in the 1950s; the predictions were too uncertain, the notion of a very early universe too remote, and the technology necessary for the experiment too exploratory. By the 1960s the technology was available, but the early universe still seemed inaccessible.

Radiation left over from the early universe, more precisely from the moment of atom formation, was discovered by accident. In 1964 Arno Penzias and Robert Wilson, two young engineers from Bell Labs, were trying to track down the cause of radio noise coming from the center of our galaxy. They started calibrating their antenna by pointing it toward the quiet night sky, at 90 degrees to the galactic center. A low-level signal appeared and persisted no matter where they aimed their antenna.

The radiation was not coming from the Earth, the Sun, or distant stars. It didn't vary with time of day or with shifts in direction. Whatever it was, the origin seemed to be the sky itself. Believing their signal was a kind of static radio noise, perhaps a calibration feature, Penzias and Wilson checked everything, dismantling and reassembling their detectors. In considering the possibility that pigeon droppings had altered the antenna's properties, they scrubbed it down; no matter what they did, the noise was unchanged. In the end they simply classified it as a signal of unknown origin.

Radio engineers tune their antennas at a given frequency. Aiming them at a target in the sky, they measure radiation intensity. The signal intensity at a given frequency from a given source is recorded as having an "equivalent temperature." The meaning of this term goes back to the discussion of Wien's Law in chapter 3. As I said there, a system in thermal equilibrium emits radiation characterized by a peak in the curve of radiation intensity versus frequency, a peak fixed by temperature. What I didn't say in chapter 3, and which I hope is clear from the figure on page 114, is that there are an infinite number of such radiation-intensity-versus-frequency curves, each having a peak at a different temperature. Very importantly, no two of these curves *ever* intersect. In other words, a measured intensity at a given frequency, one point in the space of all curves, lies on one, and only one, of those curves. To see this, look back at the plot of radiation intensity versus frequency in chapter 3.

Now comes the confusing part. Suppose the engineer fixes the frequency for the reading and then proceeds to find the source's intensity. The result is a frequency-intensity pair. If the source is in thermal equilibrium, the measurement yields the source's temperature. But even if the source is not in thermal equilibrium, the measurement still necessarily falls on one and only one of those curves. In the first case the curve tells the source's temperature. In the second case it doesn't; it gives a reading of its "equivalent temperature," simply a label of which curve the point falls on. In Penzias and Wilson's case, the equivalent temperature was 3 degrees Kelvin.

The uniformity of the signal and the absence of any source strongly suggested a cosmic origin. The sky seemed to be uniformly at an equivalent temperature of 3 degrees. However, it made no sense to think of the empty night sky being in thermal equilibrium. A thermometer in outer space would certainly not register 3 degrees. Fortunately, a solution to the puzzle was put forth immediately. A young Princeton University theoretical astrophysicist named James Peebles had been thinking about what kind of uniform radiation might be visible in the night sky as a remnant of the thermal equilibrium conditions that held early in the universe's history.

One of the most interesting features of the radiation-intensity-versus-frequency plot is the striking resemblance it bears to the figure of the number-density-versus-energy plot proposed by Maxwell for molecules in thermal equilibrium. This is no accident. As Einstein first showed and I'll discuss in detail in chapter 6, radiation can be thought of as made up of photons, or *quanta*—packets of energy. A photon's energy is proportional to the frequency of the radiation, and the intensity of the radiation depends on the number of photons. Viewed this way, intensity versus frequency is very much like photon number versus photon energy, and, not surprisingly, radiation in thermal equilibrium will also have a peak at a value fixed by temperature.

In order for radiation to be in thermal equilibrium, photons need to interact with their surroundings. They are sensitive to electric charge, so they scatter readily off negatively charged electrons and positively charged protons, both present in the dense pre-3000-degree environment of the early universe. Continually readjusting their motion by these scatterings, those early photons maintained thermal equilibrium and a common temperature with the electrons and protons, even though the photons, being electrically neutral themselves, did not scatter directly off of each other. However, once hydrogen atoms formed, the situation changed; the bound combination of an electron and a proton is electrically neutral and therefore essentially invisible to photons. Photons ceased interacting with their surroundings once atoms formed, but they didn't disappear.

The photons Penzias and Wilson detected and the ones we still see today have been moving unperturbed through the universe for the last fifteen billion years. Unscattered, the photons have nevertheless changed in one significant way because of the universe's expansion. Space's stretching means the characteristic distance between any two points has grown. The growth factor, the increase in wavelength, is approximately one thousand. This means that that the wavelength of a photon in thermal equilibrium when atoms formed has become a thousand times larger over the past fifteen billion years.

The key is that all the photons grew the same way, so their relation to one another in the distribution is unaltered. Wien's Law has another interesting feature: a thousandfold increase in wavelength, or, equivalently, a thousandfold drop in frequency, yields the same curve as a thousandfold drop in

Arno Penzias (right) and Robert Wilson in front of the antenna they used to measure the cosmic microwave background radiation

temperature. Since the location of the frequency peak is proportional to temperature, Peebles showed that shifting that peak by a factor of a thousand takes you to a plot shifted in temperature by a factor of a thousand. In describing the photons Penzias and Wilson saw, the technically correct phrase is "they came from a 3000-degree thermal distribution stretched in wavelength (contracted in frequency) by a factor of a thousand." The much simpler phrase is "3-degree photons."

The Penzias-Wilson measurement was only one point on a curve of intensity versus frequency, a suggestive one with a convincing theory backing the hunch. It took another twenty-five years to clinch the argument, twenty-five years until measurements at all frequencies traced out the full thermal equilibrium curve, stretched of course by a factor of a thousand. Contradictory results abounded during that quarter century, doubts were raised, alternative explanations found. The high-frequency part, absorbed in the Earth's atmosphere, is inaccessible to ordinary telescopes. To circumvent this difficulty, recording instruments were placed in balloons and flown, and antennas were sited on mountaintops. It wasn't enough. It became increasingly clear that the issue was not going to be settled until the graph could be studied in its entirety

COBE, the acronym for Cosmic Background Explorer satellite, was designed to settle once and for all the issue of the cosmic microwave background's spectrum. In January 1990, two months after the launching, COBE's project leader gave a talk at the American Physical Society. John Mather's data showed beyond a shadow of a doubt the full curve of radiation in thermal equilibrium. No questions any more. Twenty-five years earlier it had been an unexplained noise. It was now simply 2.735 degrees Kelvin radiation. The audience rose and gave Mather a standing ovation.

COBE carefully traced out the 2.735-degree curve, focusing on a particular direction in space. Another detector on COBE was designed to move easily across the sky, sampling the radiation at different angles. The second detector was searching for small but statistically significant differences in the radiation's temperature, 2.735 in one direction and perhaps 2.736 in another. The idea behind this was simple: the universe could not have been perfectly uniform when atoms formed. Some small seeds must have already been there, seeds that grew with time to become the fluctuations that produced galaxies. This second set of COBE data took longer to collect and analyze than the single-angle evidence. The graph presented at a late April 1992 meeting of the American Physical Society plotted square-of-temperature difference, measured in millionths of a degree Kelvin, versus the angular separation of the readings' directions. The expected small differences were there! By now all introductory cosmology courses show a temperature map of the universe.

I can't leave my discussion of the early universe without coming back one more time to neutrinos, my tool in trade. You know what happened to the photons, the protons, the electrons, and the neutrons, but there were also neutrinos in thermal equilibrium with them at a hundred thousand million degrees. Penzias and Wilson measured the photons; there are several thousand of them in every cubic inch of the universe. But there are also several thousand neutrinos in every cubic inch, about a hundred million of them for every atom in the universe.

The neutrinos are almost certainly there, distributed in a thermal distribution at 2 degrees. They are not really at 2 degrees: that's just their equivalent temperature, the reflection of the stretching of space since the last time they were in thermal equilibrium with surrounding matter. The temperature was then ten billion degrees. Nobody has ever seen a single one of those neutrinos, though I and many others have thought about ways to do it. Perhaps it will take a hundred years before we can trace out their curve and check if it's thermal equilibrium at 2 degrees. Perhaps it will be even longer. Maybe it won't even be a curve of thermal equilibrium at 2 degrees. If it's not, we will have to revise our notions of the early universe. We will then be on to new mysteries.

In a way, it seems odd to spend so much time, effort, and money to look for neutrinos that have not interacted with anything since a second after the Big Bang. Seen from another point of view, our own existence is ephemeral: less than two million years since Homo erectus emerged. It's been a little longer for primates, a few hundred million for multicelled organisms, a few billion for life, Earth, and the Sun. Those neutrinos have been moving undisturbed for fifteen billion years. We don't know for sure yet, but they are probably more important to the future of the universe, its expansion or contraction, than all the stars and all the planets. They are certainly far more important to the universe than we are.

The Big Bang and the Big Crunch

The observation of the 3-degree radiation in the mid-1960s made the scientific community realize that the early universe's complexities might be deciphered. That confidence grew rapidly with the prediction of the hydrogen-to-helium ratio and has continued ever since. Finding with greater precision the subtle patterns of temperature differences at different angles is now the frontier. Special telescopes, balloons, and satellites are looking for them. New instruments under construction will allow us to look further back in time. A hundredth of a second after the Big Bang, a hundred billion degrees Kelvin, is thought of as the universe's advanced middle age. Researchers want to know about infancy; they want to run the full temperature gamut—from high temperatures to higher ones, possibly even all the way back to infinite temperature.

Will the universe continue expanding or will it fall back on itself toward a future Big Crunch? Will it continue cooling or is it headed for a Big Roast? Even though such a conflagration would not occur until more than fifteen billion years from now, we'll soon know what the universe's ultimate fate will be. Some rockets fly skyward and never return, while others perform a graceful arch up and back. Determining which of the two trajectories the rocket will follow depends on two factors: the rocket's initial velocity and the mass of the

Earth. Likewise, velocity of expansion and average mass density set our universe's future. The first is known and the second will be determined within a decade by experiment.

The favored cosmology model at the moment is one that minimally allows us to avoid the Big Roast. We are the analogue of the slowest possible rocket that can escape the Earth's gravitational field. Phrasing the issue in cosmological terms, the universe is in an eternal outward gliding.

Visible matter—stars and dust—makes up only a few percent of the necessary material to achieve this state, known as a "flat universe." Some other kind of matter, whose identity is still a major mystery, places on the scales another 30 percent of what is needed. This matter is not an emitter of light and therefore remains undetected by conventional telescopes. The remainder, more than 60 percent, is provided by the cosmological constant, a kind of energy originally proposed by Einstein to account for the then-favored model of a stationary universe, neither expanding nor contracting. When later evidence proved the universe was expanding, Einstein did away with the cosmological constant. Surprisingly, we now seem to need it back, with just the right magnitude for the very special kind of expansion known as flatness. Some future insight may mean we can dispense with it again, but I wouldn't bet money on it. If it is present, the universe's expansion is accelerating.

This describes our universe, but still doesn't say how it started on its voyage of expansion. The embarkation poses some interesting temperature problems. Photons reach us from all directions after their fifteen-billion-year journey, carrying imprinted on them their place in the 2.735-degree distribution. What or perhaps who told the universe to start with the same temperature everywhere? Significantly, although the universe started at one instance of time, it did not start in one point. It began everywhere, and space itself has been stretching ever since then. The photons to the right and to the left are meeting for the first time. If so, and if they have never been in contact before, why are they at the same temperature?

The most widely accepted solution was proposed twenty years ago by a physicist named Alan Guth. He envisioned a very early phase in which an infinitesimally small region underwent an extraordinarily rapid expansion: Guth called it inflation. At the end of that growth the heated universe settled into its normal mode of expansion, retaining, however, the coordinated memory of the bubble it was born in. The condition of equal temperature to the left and to the right, set in the early phase, remained forever untouched.

The model has competitors, most of them variations on its theme. That doesn't necessarily make scientists happy. They like nothing better than a challenge, a good fight to sharpen model building and provide experiment with a target to aim at. Experiments sometimes disprove theories, occasionally confirm them, and often simply lead to new puzzles. Competition, as economists keep emphasizing, is good. There's also the safety issue. Michael Turner, a well-known University of Chicago astrophysicist, recently said, "We'd like to have a backup—a worthy competitor—something in our back pocket in case inflation doesn't work out."

Some astrophysicists may have very recently come up with a worthy competitor. The model of Justin Khoury, Burt Ovrut, Paul Steinhardt, and Neil Turok requires at least one additional dimension to the familiar four—three spatial and one for time. Of course, any extra dimensions have to be hidden well enough to have escaped detection. One way for this to happen is for them to be very small. Imagine an ant crawling along an inch-wide, mile-long pipe. The ant can freely travel down the pipe, but never more than an inch in a direction perpendicular to its primary motion. In describing the location of the ant, we normally specify how far along the pipe the ant has traveled but ignore its gyrations around the pipe. In such a one-dimensional description, the position relative to the pipe's central axis becomes an extra, ignored dimension. Likewise, an electron could sample deviations of a billionth of a billionth of an inch in a fifth dimension without our awareness of its departure from the straight and narrow.

The idea of extra dimensions is not new. In the 1920s, an extension of Einstein's general relativity theory to five dimensions was shown to imply the joining of electromagnetism to gravity. This result entranced Einstein, holding forth the promise of a unified theory of the two fundamental forces then known. Though that union didn't quite work, the past twenty years have seen a renaissance of models with extra dimensions.

If they are present, extra dimensions are now very small, but they may not have always been so. The newest model envisions several simultaneously existing early universes, conceived for simplicity as sheets. Four dimensions moving in a fifth is too hard to visualize. Consider motion perpendicular to the sheet. Since the sheet is a whole universe, motion perpendicular to the sheet must take place along a fifth dimension, carrying the whole universe with it. But the sheet is not alone. Other sheets are also moving in that extra dimension. An enormous amount of heat is generated when and if two sheets collide; ripples from the collision spread along each sheet and the heated universe on each sheet begins its journey of cooling and expansion. Of course, as movement in the fifth dimension continues, another sheet might be encountered, bringing an end to the universe. Hence the name "ekpyrotic universe," ekpyrosis being the model conceived by Stoic philosophers in which the universe is regularly consumed and then reconstituted by fire.

Cosmology is a field that has existed ever since our ancestors first looked up in wonder at the night sky and tried to assign a meaning to what they were seeing. The twentieth century is the century in which that wonder was converted into science. From Einstein's theory of general relativity to Hubble's discovery of an expanding universe to Penzias and Wilson's observation of the cosmic microwave background—the wonders have not ceased and much more may be in store.

The emergence of modern science can be dated to Copernicus's suggestion that the Earth is not the center of the universe. From there we have progressed to viewing our Earth as an ordinary planet circling a midsized star located toward the edge of a typical galaxy. Centuries from now, our descendants may look back and say this was the moment humankind began to realize they live in an ordinary universe surrounded by countless undetected others. Perhaps some are bigger, others smaller, some hotter, and some colder. Weinberg's *The First Three Minutes* ends with a phrase that is as good a justification as I can imagine for studying these questions: "The effort to understand the universe is one of the very few things that lifts human life a little above the level of farce, and gives it some of the grace of tragedy."

Chapter 6

THE QUANTUM LEAP

IN THE BOOK'S INTRODUCTION, I proposed temperature as a guide in exploring some of the great science ideas of the past and present. Following that thread, I started with the familiar, our own bodies, investigating the origins of fever and our similarity to other living organisms. Once imagined as unique creatures living on the central celestial body of the universe, we now think of ourselves differently. In part due to centuries of better temperature measurements, the focus of modern studies is plate tectonics, genetic sequencing, and Big Bang cosmology. Still to be addressed is chapter 6's theme, quantum mechanics, seen of course from a temperature vantage point. Profoundly reshaping our notion of reality, quantum mechanics reaches into our day-to-day lives, elucidating why computer chips work and why hydrogen unites with oxygen to create water. In another realm, quantum mechanics shows why the Sun's core heats us now and why the core won't collapse even after the Sun turns into a white dwarf star. Temperature has a profound role in all these processes.

This final chapter has two subthemes, youth and absolute zero. The first revolves around the struggle of the young and the old. People often say that mathematicians and theoretical physicists do their best work before they are thirty. The reasons given are many: vigor, ambition, greater ease in jettisoning older preconceived notions. All are true and we have seen examples of youthful scientific exploits, from Maxwell's discovery of statistical mechanics to Carnot's insight into heat engines' efficiency. Then there's the miracle of 1666, when twenty-four-year-old Isaac Newton discovered both calculus and the inverse square law for gravity. Equally impressive is the 1905 accomplishment of twenty-six-year-old Albert Einstein in formulating the Special Theory of Relativity and laying the groundwork for the new quantum physics.

Youthful inspiration is also striking in the 1926 birth of quantum mechanics, marked by the rapid ascension to intellectual leadership of Paul Dirac, Enrico Fermi, Werner Heisenberg, and Wolfgang Pauli, all twenty-five or younger at the time. But the story, as with most human enterprises, is never quite that simple. Niels Bohr, forty-one in 1926, was the guiding spirit of quantum mechanics' interpretation. Ein-

Subrahmanyan Chandrasekhar at age twenty-four, at about the time he provided the definitive evidence for the Chandrasekhar limit for the collapse of white dwarfs

stein, forty-seven in the year of quantum mechanics' formulation, remained a diligent critic of the theory he had done so much to shape. So it's not only a young person's field, but the young do tend to dominate. If you are not impressed by twenty-three-year-old Heisenberg discovering a form of quantum mechanics, consider the chapter's closing section, with nineteen-year-old Subrahmanyan Chandrasekhar's 1930 application of relativistic quantum mechanics to collapsing stars. Not too many teenagers do Nobel Prize–level work.

The chapter's other subtheme is the attempt to reach absolute zero. A feat of scientific exploration, it gained even greater importance in the twentieth century with the realization that the low-temperature world is inextricably linked to the quantum one. Superconductivity, superfluidity, Bose-Einstein condensation, and other puzzling behaviors near absolute zero can only be understood in light of the rules imposed on atoms by quantum mechanics. In turn, those lowest temperatures illuminate the quantum world.

Scientific efforts to reach lower temperatures began well before the concept of absolute zero was formulated. In 1800 the quest was phrased as an effort to liquefy all known gases. By the mid-nineteenth century, when it was realized that –273 Celsius was the lowest possible temperature and not just another point on the scale, the attempt to reach lower temperatures gained a deeper meaning. In the twentieth century, many of the quantum world's new riches were revealed only near that absolute zero.

Faraday's Perfect Gases

By the end of the eighteenth century, chemical separations had succeeded in isolating many of the components in ordinary air. The question then arose of what it might take to change them into liquids. The great Frenchman Antoine Laurent Lavoisier was the first to view in a modern way the nature of reactions like those that allow hydrogen and oxygen to form water. He understood at some level the individuality of elements and even conjectured in his *Oeuvres* that, if the temperature was lowered, the gases in the air would liquefy. This line of thought was pursued by John Dalton, whom we already met in connection with his work on gas properties and as the founder of the new school of scientists in emerging Manchester. In 1801 Dalton said, "There can scarcely be a doubt entertained respecting the reducibility of all elastic fluids into liquids; and we ought not to despair of effecting it in low temperatures and by strong pressures exerted on the unmixed gases."

I take this statement as the beginning of the race to liquefy all known gases, a race that effectively ended a little over a hundred years later with Kamerlingh Onnes's 1908 liquefaction of helium. There were many who contributed along the way: Pictet in Geneva and Cailletet in Paris, who first liquefied a few drops of oxygen on the same day in 1877; and Olsewski and Wroblewski in Cracow, who liquefied comparatively large amounts of oxygen. One of the great tools in cooling, more commonly known as a thermos bottle, is the "Dewar flask," a double-walled container named after its inventor, James Dewar, the man who in 1898 became the first to liquefy hydrogen. Shortly thereafter he even produced solid hydrogen. The story is a long and interesting one, but I'll only describe the feats of the first and the last of the great experimenters who reached for lower temperatures to liquefy gases.

The first was Michael Faraday, perhaps the nineteenth century's premier experimentalist. He spent his entire career in London at the Royal Institution, a research center founded by Benjamin Thompson, Count Rumford. We encountered

Rumford earlier as the measurer of the mechanical equivalent of heat. During his 1798–1802 stay in Britain, between his service in Bavaria and his angry departure for France, Rumford established the Institution, hoping it would become a showplace for the popularization of science as well as a research laboratory. In 1801, he recruited Thomas Young and Humphry Davy as faculty members for the new establishment, thinking they would help launch his venture. The two of them were an enormous success.

Davy started the field of electrochemistry, eventually using this technique to discover the chemical elements sodium, potassium, calcium, barium, strontium, and magnesium. A practical inventor, Davy was credited with developing the miner's safety lamp. In addition, he was a distinguished poet, much admired by his colleague in rhyme Samuel Coleridge, who proclaimed that "if he [Davy] had not been the first chemist of his age, he would have been the first poet."

Well known as a mesmerizing lecturer, Davy was a voice in the wilderness to modest twenty-one-year-old Michael Faraday, born poor and apprenticed at age thirteen to a bookbinder. Given tickets to a series of four Davy lectures by one of the shop's patrons, Faraday found the talks transfixing. He took careful illustrated notes, bound them, and then sent them off to Davy, inquiring if he could be hired as an assistant. As Faraday wrote later in life,

> My desire to escape from trade, which I thought vicious and selfish, and to enter into the service of Science, which I imagined made its pursuers amiable and liberal, induced me at last to take the bold and simple step of writing to Sir H. Davy expressing my wishes and a hope that, if an opportunity came in his way, he would favor my views: at the same time I sent the notes I had taken of his lectures.

Impressed by young Faraday's notes, Davy interviewed him and hired him six months later when an opening came up. On March 1, 1813, Faraday began work at the Royal Institution. He remained there for the next forty-five years, living most of that time in modest accommodations located in the upper stories of the building.

Faraday's experiments, simple and elegant, were carefully recorded in lab books that he bound himself, remembering the profession he had so happily left. In modern language, we would say he was a great chemist and a great physicist; Faraday described himself as a natural philosopher. With an uncanny gift for recognizing the salient points of an experiment, Faraday, while devoid of formal education, always had a sense of a larger framework. He was the guiding spirit in showing the connection between currents and magnetic fields, laying the foundations for the modern electrical power industry. Aiming for a universal theory of forces, Faraday established links between electricity and magnetism and then between the two and visible light. In many ways he anticipated Maxwell's theories of electromagnetism and the subsequent wave theory.

Faraday's work in chemistry was equally rich. He discovered benzene and isobutene, found the formula for napthalene, established the existence of the hydrocarbon family, and did pioneer work in electrolysis, steel alloys, and clays. Faraday was also the master at liquefying gases; his successes with chlorine, ammonia, carbon dioxide, sulfur dioxide, nitrous oxide, hydrogen sulfide, cyanogen, and ethylene are all recorded in his bound lab books. By 1845 he realized that oxygen, nitrogen, and hydrogen could simply not be liquefied with the tools available to him for lowering temperature and increasing pressure. Faraday, calling these and others like them "permanent gases," left their liquefaction as a problem for future scientists.

By 1898 all the gases on Faraday's list had been liquefied. Only one gas remained unliquefied, not on the original list because it wasn't detected until 1868 and then only on the Sun. The gas was discovered using the spectral line technique, very new at the time. Its name is helium, from the Greek for sun, *helios*.

The spectral line technique, which played such an important role in understanding atomic theory and quantum physics, was new in 1868, but its roots go back 200 years. In 1669 twenty-seven-year-old Isaac Newton conducted an experiment. He was going to settle the question of whether a prism adds color to sunlight—Descartes's view—or simply spreads it out. Newton took two prisms separated by a large screen with a slit in it. Sunlight coming through the first prism displayed the full range of colors. The slit in the otherwise opaque screen ensured that only single-color light reached the second prism. Regardless of which color Newton selected, transit through the second prism left it unchanged. With that, he proved the prism only spreads sunlight out into its constituent colors. As Keats later said, Newton "unweaved the rainbow."

In the early nineteenth century, Joseph Fraunhofer, with finer instruments than Newton ever had, did more than spread the sunlight. By then it was understood that each location, each different color, is another light frequency in the continuous superposition of frequencies that form the spectrum. Fraunhofer found thousands of missing frequencies, thin dark lines in the solar spectrum. Too thin to have been observed by Newton, these lines remained a simple curiosity until 1860. In that year Gustav Kirchoff decided to recreate the solar surface's temperature on a lab bench using the new burner developed by his Heidelberg colleague, Robert von Bunsen.

Heating a variety of chemical salts in the burner's intense flame, Kirchoff passed white light through the vapor formed by the salts. Each time, the emerging spectrum contained dark lines, presumably salt-absorbed light frequencies. Curiously enough, direct light from the flaming salt showed a spectrum with bright lines located at exactly the same frequencies as the dark lines of the previous experiment.

The conclusion was inescapable: heated salts can either emit or absorb light of given frequencies. Since the location of the lines varies from salt to salt, the Bunsen burner became a superb instrument for analytical chemistry. Those frequencies, those lines, different for each element, are positioned at values obeying curious formulas. Their meaning was understood fifty years later when Bohr created an accurate model of the atom.

The same lines seen in the Bunsen burner are present in the light spectrum of both the Sun and nearby stars. This confirmed the view that the Sun was an ordinary star, no different in chemical composition than many others shining in the night sky. There was also the magnificent discovery of helium, corresponding to a set of solar spectrum lines unlike anything seen in the laboratory.

The Last Liquid

In 1895, while studying heated samples of gas released by uranium ores, William Ramsay noticed that the emitted radiation had the same spectral lines as helium. This settled the debate of whether this element existed on Earth or not. A little checking told him that helium was, as expected, a noble gas, an element that doesn't combine chemically with others. Once helium was isolated, it became a new target for the program of liquefying all gases. Accomplishing this final feat involved major technical challenges, with some very surprising and important ramifications, but before discussing those, I'll take a brief side trip into the strange connection between helium and the temperature inside the Earth. The link is radioactivity.

As I described earlier, radioactivity was hailed as the new source of heat for the Sun, the provider of the energy needed to resolve the conflict between Darwin's estimate of the Earth's age and Kelvin and Helmholtz's of the Sun's. Oddly enough, radioactivity has turned out to be a much more critical source, proportionally speaking, of heat generated within the Earth than within the Sun. The fusion of hydrogen to helium and, to a lesser extent, other fusion reactions generate the overwhelming majority of solar energy. These reactions, requiring core temperatures of millions of degrees, are, however, impossible on Earth except in laboratory settings. Radioactivity, on the other hand, occurs spontaneously at any temperature, so it can take place anywhere radioactive elements are found.

People had been aware that heat is generated within the Earth long before radioactivity's discovery. Volcanoes are the obvious manifestation, but the source of that energy was unknown until the twentieth century. We now realize the two main causes, comparable in magnitude, are radioactivity and stored heat from collisions with early asteroids. Neither one alone is significant enough to account for plate movements, volcanic activity, hydrothermal vents, and the other manifestations of internal thermal activity, but together they explain it all.

Radioactivity was discovered in 1896. Interestingly, it was found in the same fairly unusual type of ore that yield helium. This seemed like more than a coincidence. Understanding how radioactive decays lead to helium requires a bit of background. To begin with, how can a chemical element such as potassium, iodine, or hydrogen be both stable and unstable against radioactive decay? The resolution lies in a fine point of nuclear physics worth going into because of its connection to temperature.

Each element of the periodic table is specified by the number of protons and of electrons in its atom. Electrical neutrality of atoms constrains the number of positively charged protons to be matched by an equal number of negatively charged electrons, but it doesn't specify how many electrically neutral neutrons can join the protons inside each nucleus. Varying the number of neutrons changes the weight of the nucleus but not the atom's chemical properties, fixed almost entirely by the orbiting electrons. Any element can appear in several chemically identical forms, differing only in neutron number. Those different forms of each element are called isotopes, conventionally distinguished from one another by the total number of protons and neutrons within a nucleus. For instance, hydrogen exists in three forms: hydrogen-1, hydrogen-2, and hydrogen-3, all having a single proton but containing, respectively, zero, one, and two neutrons. The three are chemically identical, but only the first two forms are stable. Without going into the details of the interactions within the nucleus, it is sufficient to say that in most cases one or more isotopes are unstable while the others are not. The instability is characterized by half-life, the time frame in which half of any radioactive sample will decay, comparatively short for a very unstable element and long for an almost stable one.

The combination of chemical selectivity and temporal instability is what makes isotopes so useful. In medicine, since all forms of iodine are drawn to the thyroid gland, some treatments dictate that a dose of unstable iodine-131 be mixed in with the stable iodine-127. The unstable isotope migrates to the thyroid just as quickly as the stable one; decaying with a half-life of eight days, iodine-131 releases energy into the surrounding thyroid tissue.

To date a once living object, you find the proportion of unstable carbon-14 to stable carbon-12. A living organism takes both up in a fixed ratio; after death, carbon-12 remains unchanged, but carbon-14 decays with a half-life of 5730 years. The ratio of the two tells when the organism died.

Potassium-40, thorium-232, uranium-235, and uranium-238 are the four radioactive isotopes that contribute significantly to the Earth's heat. The first decays into calcium and the last three into isotopes of lead. Other radioactive elements and other isotopes may have been more important on the early Earth, but their comparatively short half-lives, significantly less than the age of the Earth, means they have already decayed away. They contributed to a more violent level of geological activity through a heat production by radioactivity perhaps five times as high as now. The four isotopes I listed are, however, key to heating the present-day Earth because their half-lives range from a little under a billion to a little over ten billion years—not so long that they don't contribute heat to the Earth now and not so short as to have already decayed away.

To appreciate the power of that radioactive heating, think of the plight of miners. For hundreds of years, miners have noticed temperatures rising as they have gone underground; jackets and shirts have been shed in the search for the deep veins of ore. The deepest mine in the world is the Western Deeps gold mine near Johannesburg in South Africa; the shaft descends a little over two miles. The bottom of the South African gold mine is a steady 140 degrees Fahrenheit; workers need to have cold air pumped down to survive.

Deeper holes have been drilled, though not for mining purposes. The present record is the eight-mile-deep Kola Hole in the former Soviet Union. The U.S. record is a hole about six miles deep made by Oklahoma gas drillers in the 1970s. However, these holes are plagued by troubles because, by five miles down, the temperature is almost 500 degrees Fahrenheit, hot enough to damage most drilling equipment. Fifty miles down, as I said earlier, the temperature reaches thousands of degrees. The quick temperature rise, 400 degrees Fahrenheit in five miles, has a simple explanation. Potassium, thorium, and uranium are mainly concentrated in the Earth's outer crust. The Earth's temperature heading toward the core continues to rise, but the greatest temperature gradient lies in those first few miles, where radioactive heating is at full strength.

Thorium-232, uranium-235, and uranium-238 all decay by the emission of alpha particles. Rutherford, with his usual great intuition, began to suspect that alpha particles were helium nuclei. By 1904 he was almost sure of it, and by 1908, he proved it. In leaving the heavier parent nucleus, the helium nuclei gather up electrons through collisions with other atoms and molecules, turning into full-fledged helium atoms. That helium, being light and inert, tends to simply float away into space, explaining its comparative rarity on Earth. However, the source is constantly replenished by new radioactive decay. The heat in the gold mine and the helium scientists were trying to liquefy have the same source, radioactivity.

Helium became the ultimate challenge in the race to liquefy all gases because it remains a gas past the point where all others, including hydrogen, have already turned to liquids. The competitors lined up for the grand finale of the race. The jump-off point was liquid hydrogen temperatures. From there, with judicious refinements, it was hoped that helium would quickly follow. As 10 degrees Kelvin was reached without any success, the contenders realized just how hard it was going to be. James Dewar, Faraday's successor at the Royal Institution, had liquefied hydrogen, making him the early favorite. In the end, the man who succeeded was a new entry from Holland, Heike Kamerlingh Onnes.

Onnes and Rutherford are the two great wellsprings of twentieth-century experimental physics, the respective founders of the modern schools of low-temperature physics and nuclear physics. In their own ways they are also transitional figures, two men who mark the change from an era in which a solitary worker built his own apparatus, made his measurements, and perhaps even bound his laboratory books, to the modern world where a team is assembled to work with specially designed equipment and the results are reported in specialized professional journals. Onnes and Rutherford each received a Nobel Prize, Onnes for liquefying helium and Rutherford for deciphering the chain of decays in radioactivity, but each made a later discovery that was much more far-reaching. Onnes's was superconductivity and Rutherford's was the atomic nucleus.

Rutherford at Manchester and later as head of the Cavendish Lab at Cambridge gathered students, research assistants, and visitors around him. Onnes built a group and a research empire in Leiden. He may also have been the first scientist to realize that modern research was going to require specially trained technicians to build and operate complex apparatus. In 1901 he founded the Leidse Instrumentmakers-school (Leiden School for Instrument Makers), where young men were and still are trained to work in scientific laboratories. Onnes opened his door to visitors, lent out equipment, forged ties to industry, and set up a special journal to regularly report his laboratory's results.

The organizational buildup was slow and methodical. Technical delays sprang up. The citizens of Leiden closed down the enterprise when they discovered Onnes was using compressed hydrogen, a very explosive gas. During the early-nineteenth-century French occupation of the Netherlands, an ammunition ship docked on a canal had blown up, destroying a large part of central Leiden. Onnes's lab was built on those ruins. The citizenry understandably didn't want to see history repeated, especially by one of their own. It took two years for Onnes to reassure his fellow Leideners. Delay followed delay, but his competitors did not beat him, because liquefying helium turned out to be harder than anyone suspected. When Onnes was finally ready to make his own push, he had all the needed tools.

Onnes's lab liquefied hydrogen in 1906, eight years after Dewar had first done it. When they succeeded, however, their large-scale enterprise immediately produced almost four quarts of liquid hydrogen a day, an enormous amount compared to Dewar's few drops. With plentiful supplies of liquid air and liquid hydrogen, the final plans to liquefy helium were set in motion in June 1908. The experiment began on July 10 at 5:45 A.M. With twenty quarts of liquid hydrogen on hand by early afternoon, the experimenters loaded helium gas into the liquefier and its temperature began to drop. Cooling continued for the next hours and then stalled, but no liquid was seen adhering to the sides of the glass beaker. Finally at 7 P.M., when the liquid hydrogen coolant was almost all used up, a light placed beneath the flask showed the reflection from fluid at the bottom of the flask. Helium had finally been liquefied. At 10 P.M., the experiment was concluded and everybody went to bed, probably after a little celebration.

This marked the end of the quest to liquefy all known gases. The challenge had been met. The event was significant enough to warrant a Nobel Prize for Onnes. The deeper importance, not obvious at the time, lay in reaching temperatures low enough for many new phenomena of the quantum world to manifest themselves. It's almost as if, after focusing for many years on reaching the highest peak in a mountain range, one discovers the truly important goal is the view from the summit of a whole new land that lies beyond.

Incidentally, as a comment on the subtheme of young versus old, I hope it's also clear that theoretical and experimental physics are very different. A twenty-five-year-old Newton or Maxwell can have a brilliant idea and do something great very quickly, but the methodical building-up of techniques, material, and personnel for experimental work doesn't take place in a flash. Rutherford was forty-three when he discovered the atomic nucleus and Onnes was fifty-five when he liquefied helium.

Superconductivity

Once they had enough liquid helium to cool other materials to a few degrees above absolute zero, the Leiden group began studying how matter behaves at very low temperatures. Resistivity to the flow of electric current or its inverse, conductivity, was an obvious choice since accurate measurements of electric currents are easily made.

Resistivity is known to decrease and conductivity to increase as temperature drops. However, nobody had ever studied these properties at liquid helium temperatures. The first set of measurements from Onnes's lab showed the expected decrease. Onnes believed this trend would continue until absolute zero. This wasn't a foregone conclusion; Lord Kelvin conjectured that electrons, the carriers of the electrical current, would slow down and eventually stop as the temperature approached absolute zero. This meant current would cease to flow or, equivalently, resistivity would approach infinity as temperature went to zero.

Onnes and his coworkers started their tests with platinum and gold, known to be good conductors. Thinking that any impurities in the metals might act to block the current's flow, they next turned to mercury, a liquid at room temperatures, making it relatively easy to purify by repeated distillation. By the fall of 1911, Onnes and his team were ready to make conductivity measurements in very pure mercury at 4 degrees above absolute zero.

The team running the experiment consisted of Onnes, Gerrit Flim, Gilles Holst, and Cornelius Dorsman. Onnes and Flim regulated the cooling apparatus next to the mercury wire. Holst and Dorsman sat in a darkened room 150 feet away monitoring the resistivity with a special current-measuring device called a galvanometer. The needle, which indicated changes in current and therefore changes in resistivity, had a light beam mounted on it so any small deflections of the needle could be seen easily. Pure liquid mercury was placed in a thin, U-shaped glass capillary tube; current flowed in and out of the mercury through platinum electrodes attached to the ends. When everything was set, Onnes and Flim began to lower the wire's temperature. As expected, resistivity decreased even though the mercury in the capillary froze solid. Then came the unexpected. At 4.19 degrees Kelvin, the wire's resistivity suddenly dropped to zero. The set-up was taken apart and reassembled. Same result. The process was repeated several times over the following weeks, always a slow and laborious operation because of the difficulties in preparing the liquid helium necessary to cool the wire. The answer never changed.

Onnes decided to put the mercury in a thin, W-shaped

Heike Kamerlingh Onnes in his low-temperature laboratory in Leiden

tube with electrodes at the kinks of the W as well as at both ends, allowing the team to check simultaneously four separate mercury wire segments. There might be a short-circuit somewhere, but the resistivity of all four pieces of wire shouldn't go to zero simultaneously. They did. At a loss, they kept on trying, repeating the experiment again and again. Finally, during one run, Holst saw the light beam mounted on the control room's galvanometer needle suddenly swing back, indicating resistivity had returned. Nothing had been changed, or so they thought.

They soon discovered that one of the students from the School for Instrument Makers had dozed off. His job was to make sure no precious liquid helium escaped from the vessel that held it. This was done by keeping the inside vapor pressure container slightly below atmospheric pressure. If a tiny leak developed, air would flow in, freeze, and seal the leak before helium was lost. The student regulated the vessel's pressure; he supposedly was constantly alert, carrying out the monotonous job hour after hour. The experiment had gone awry because the student had nodded off, a small leak had developed, the temperature rose, and the resistivity returned. Human error, not science, had skewed the measurement.

Repeated experiments convinced the group the mercury wire's resistivity dropped precipitously to zero at 4.19 degrees Kelvin and came back just as suddenly when the temperature was raised above 4.19. Other metals were tried: tin and lead wires also had resistivities that, though sizable at liquid hydrogen temperatures, vanished at very low temperatures. The phenomenon of vanishing resistivity is called superconductivity.

Onnes hoped this new discovery would have important practical applications, but didn't live to see them realized. The advantage of using superconducting wires for power transfer is compelling. As already mentioned in chapter 2, while discussing James Joule's work proving the interchangeability of different forms of energy, heat is generated in passing an electric current through a wire. Joule heat, as it is now known, is proportional to the resistivity and to the square of current intensity in the wire. In some appliances, such as a toaster, it's desirable to have electrical energy converted to heat, but transmission with as little loss of energy as possible is the usual aim.

In transferring electrical power from a hydroelectric dam to consumers hundreds of miles away, low resistivity is the goal. Since energy dissipated through heat loss is wasted, this amounts to a major expense in electric power production. Superconducting wires would cut that loss to almost zero. Furthermore, much more current can be pushed through a superconducting wire without fear of damage. In a trial, a company called American Superconductor is currently replacing 18,000 pounds of 1930s-vintage copper wire in a Detroit Edison Power station with 250 pounds of superconducting wire. The cost/efficiency saving is substantial.

The semiconductor has transformed technology in the second half of the twentieth century, making possible the transistor, the integrated circuit, and the computer. The superconductor could change the cost of electrical power and the way it is distributed. Magnetic resonance imaging (MRI) units already use superconducting magnets, created by wrapping superconducting electric coils around cores. The large currents needed to produce the strong MRI magnetic fields would have too much Joule heat if these wires weren't used. Trains that operate by superconducting magnetic levitation are being tested. But there's a catch to all this promise: temperature.

It's both hard and expensive to reach the very low temperatures necessary for superconductivity. While a hundred-mile-long cable of superconducting tin may lose less energy than a copper wire, the advantage is academic if the cable has to be packed in hard-to-produce and hard-to-maintain liquid helium. Twentieth-century research on superconductivity has been a major thrust in physics, chemistry, and the emerging new field of materials science. The two goals have been understanding the mechanisms that make superconductivity work and finding new and better materials, namely, ones that superconduct at higher temperatures.

Superconductivity continues to amaze us, but progress has been slow and arduous. Researchers have tested all kinds of metals and alloys, looking for materials that show superconductivity at higher temperatures. From a commercial point of view, the break-even point is 77 degrees Kelvin, the temperature at which nitrogen turns liquid. Liquid nitrogen is easy to make and cheap, typically a hundred times less expensive than liquid helium. Keeping power wires at 77 degrees Kelvin is feasible even on a large scale. Keeping them at 4 or even 10 degrees is not.

Until the mid-1980s, no material had been observed with superconducting properties above 30 degrees Kelvin, still far from liquid nitrogen temperatures. A bombshell hit the field in 1986, a double whammy of sorts. Georg Bednorz and Alexander Muller, two scientists at IBM's lab in Zurich, found that a ceramic-like substance, LBCO (lanthanum barium copper oxide), was superconducting at 35 degrees Kelvin. The new material was very different from ordinary metals, holding the promise of entirely new kinds of superconductors. The excitement became a frenzy a year later when Paul Chu and Maw-Kuen Wu announced that YBCO (yttrium barium copper oxide) was superconducting at 93 degrees, well past the 77-degree liquid nitrogen barrier. Suddenly the sky was the limit. People started talking about room-temperature superconductors and new technologies that would change our style of living. In 1993 MBCCO (mercury barium calcium copper oxide) was found to be superconducting at 134 degrees at ordinary pressures and at even higher temperatures if squeezed.

More than a decade later, all one can say is it hasn't been

that easy. These new materials, the copper oxides, aren't like metals. Many of them only conduct at all in peculiar chemical configurations. They are sensitive to stray magnetic fields and to impurities. Ceramic in nature, they are brittle, hard to shape, and very difficult to roll into wires. It doesn't mean it can't be done. The stakes are high and the research intense, but we still don't know if the copper oxides are the magic carpets we dreamed of. One thing was clear almost from the start with these new materials. Their form of superconductivity is not like that of metallic superconductors. Resistivity suddenly drops to zero, the hallmark of superconductivity, but other features of the transition are different. At a fundamental level, nature seems to have devised at least two independent solutions to the problem of how to have a current flow with zero resistivity.

Is room temperature superconductivity possible? Could there be other mechanisms we still haven't discovered? Are we likely to have more advances? Probably all three. Even as I write, there's a new excitement generated by the discovery that cheap and easy-to-prepare magnesium diboride is a metallic-type superconductor at 39 degrees Kelvin, about 10 degrees above the previous limit. We just don't know what the future is going to yield.

Onnes would have loved to see all these new developments. He died in 1926, just three weeks before his lab succeeded in solidifying helium. Hendrik Casimir, one of the greats of Dutch physics, remembers the story of Onnes's funeral:

> He was living at the outskirts of Leiden and the family had a family vault at a nearby cemetery at a distance of four kilometers or so. It had been ordained by Kamerlingh Onnes that his faithful technicians should follow the cortege on foot. It was rather a hot day and there they came

> in morning coats and top hats. As often in such cases, the procession was rather late in leaving the house and at the cemetery many officials were waiting. So, as soon as they were outside Leiden, the whip was put on the horses and they went at rather a brisk pace, with the technicians walking or trotting behind. They arrived at the cemetery sweating and puffing and then one of them looked at the others with a broad grin on his face and said, "That is just like the old man; even after his death he keeps you on the run."

The field of low-temperature physics can also be characterized as "on the run." Onnes hoped to turn liquid helium into a solid. It turns out that no matter how much you lower the temperature, helium will never solidify unless you increase the pressure as well. The Onnes lab found the new form at 2.5 degrees Kelvin when the pressure reached 25 atmospheres. New developments show solid helium changes crystalline structure at 30 atmospheres and 1.8 degrees. As for the low-temperature limit, that frontier has moved on as well. David Lee, Douglas Osheroff, and Robert Richardson shared the 1996 Nobel Prize for studying different phases of conventional helium-4's isotope, helium-3. The A phase sets in below 2.7 thousandths of a degree and the B phase below 1.8 thousandths of a degree above absolute zero.

What future findings and applications are likely to be depends on many factors, including possibly the sleeping habits of student lab technicians.

Duality, Exclusion, and Uncertainty

Understanding superconductivity or knowing why some materials are electrical conductors and others insulators requires a bit of quantum mechanics. Without quantum mechanics, the periodic table of elements is simply a set of rules for grouping similar chemicals, a convenient but puzzling zoological-like classification. With quantum mechanics, the behavior of elements is comprehensible and predictable. Without quantum mechanics, atomic behavior near absolute zero temperature is bizarre.

Quantum mechanics is a subject renowned for its difficulty, so I'm only going to discuss three important principles. Duality, exclusion, and uncertainty are odd, counterintuitive, and seemingly paradoxical. But then, quantum mechanics is not simply another way of looking at familiar concepts; it's a profound new shaping of reality. All three principles were first enunciated by twenty-five-year-olds, adding to the mystique of quantum mechanics and its reputation of being the province of wunderkinder.

In a way, it all started in 1900 as a temperature problem. By then thermodynamics was a well-established subject. The theories of heat, light, thermodynamics, and electromagnetism all stood on solid ground. Chemistry and physics had never looked better. Admittedly the frequencies of spectral lines followed odd rules, but the general emission and absorption of radiation was certainly consistent with thermodynamics. The theory of specific heat, the amount of heat needed to raise temperature by 1 degree, showed some strange features, mainly centered on the somewhat mysterious behavior of solids, but again that didn't seem to threaten any fundamental principles.

There was one striking problem in 1900 thermodynamics, so glaring it couldn't be brushed away. Wien's Law predicts a maximum in the curve of radiation intensity versus frequency. The maximum is at a value fixed by temperature, a relation we have already used many times. The problem was that all attempts to derive the shape of the curve failed. Worse than that, they all gave a nonsensical result, a curve with no maximum. Theory was predicting a continued rise in intensity as frequency grew. The prediction, disagreeing starkly with experiment, was obviously wrong.

It seemed that something was amiss with scientists' approach to the interlocking problems of emission and absorption. Thinking an idealized situation might help, physicists began to consider perfect emitters and absorbers. In studying mechanics, you begin with a perfectly smooth object sliding on a perfectly frictionless plane. Similarly, Wien's Law was examined for the case of an insulated oven with perfectly absorbing inside walls, heated to an absolutely uniform temperature and every square inch of its surface identical. With a small window on one side of the oven as the only outlet for radiation, the intensity-versus-frequency curve would be examined under ideal conditions. The emission was called blackbody radiation. Even in the tightly controlled situation, theory continued to disagree with experiment. It predicted a steady growing of intensity with increasing frequency, not the hoped-for rise and fall. The disastrous never-ending rise came to be known as the "ultraviolet catastrophe," continued growth of intensity past the violet range of frequency.

Several scientists struggled to patch the theory up, none more vigorously than Max Planck. Atoms presumably generate the radiation when stimulated by the oven's heating. The details of what initiates the radiation were shown to be unimportant. All that mattered was the notion that the energy of the generators is smoothly distributed around a temperature-fixed average value. No theory seemed to give the right answer, nothing made the curve show a peak. In 1900 Planck gave up using conventionally accepted ideas and took a bold step forward to obtain a formula that fit the data. As Planck later reminisced,

> It was an act of desperation. For six years I had struggled with the blackbody theory. I knew the problem was fundamental and I knew the answer. I had to find a theoretical explanation at any cost, except for the inviolability of the first two laws of thermodynamics.

The generators' average energy was fixed by temperature. Planck couldn't abandon that assumption. Doing so would have violated the laws of thermodynamics. He could, however, and did give up the smoothness of the distribution about the average. His proposed scenario was one in which energy could increase or decrease in steps instead of in a continuous change. He called these steps *quanta*. The amount of energy in each step was proportional to the frequency of the radiation. The constant of proportionality was a number now aptly called Planck's constant. It was as if a dispensing machine replaced the old grocer who weighs out on a scale the amount of flour you want. There are still no limits to the total you can purchase, but it is sold packaged in one-pound sacks and you pay in multiples of a single-unit currency.

Planck was a reluctant revolutionary. Young Albert Einstein was an eager one. In the same 1905 annus mirabilis in which Einstein overthrew the established notions of space, time, and simultaneity, he attacked the quantum problem with characteristic boldness. Planck said the generators of radiation in a cavity's wall had quantized energies. Einstein proclaimed this only made sense if radiation itself was quantized. Emitted, absorbed, or simply traveling, radiation energy always comes in packets, *quanta*.

Einstein changed the scientific perception of light in two ways. With his Special Theory of Relativity, he showed light travels at the maximum speed a signal can have. With his version of the quantum theory, he reshaped the very notion of what is light. In 1670 Isaac Newton said light was corpuscular. One hundred thirty years later, Thomas Young, Humphry Davy's colleague as first faculty members of the Royal Institution, proved light was a wave.

In 1905 Einstein was saying that light or any electromagnetic radiation is both. It has a dual nature. When viewed as a particle, it acts like a particle; when viewed as a wave, it acts like a wave. As a particle, it carries energy proportional to the frequency of the wave; as a wave it has frequency proportional to the particle's energy. The *particle,* subsequently known as the *photon,* is admittedly a rather strange sort of particle; it has no mass and travels only at the speed of light, the same speed as the wave. The photon is neither a particle disguised as a wave nor a wave disguised as a particle. It is both. Einstein was making physicists dizzy with his duality. The 1905 insight launched the quantum theory. Einstein was awarded the 1922 Nobel Prize in physics for it. Many felt it should have come much sooner.

Convinced of quanta's reality, Einstein rapidly proceeded to make others believe in them. In 1907 Einstein studied the vibrations of the atoms inside solids. At room temperature, heating a diamond increases its temperature by less than 20 percent of the then expected value. Einstein showed the right result follows if one assumes the underlying atomic vibrations are quantized, no longer simply taken as a smooth distribution about a temperature fixed average.

Though it took Einstein's genius to show the right path, the problem of explaining the heating of a diamond was almost straightforward compared to that of understanding the frequencies of lines in light spectra, but that was also quickly solved. In 1913 Niels Bohr invented a new model, according to which the atom has a central nucleus surrounded by orbiting electrons whose energies are fixed by quantum rules. The combination of Bohr's model and Einstein's version of quantization gave the answer to the surprising regularities in the frequencies of radiation observed with spectral lines. Heating a substance to high temperature makes the electrons hop into higher orbits. They later drop back down, emitting a photon, giving to that photon the energy the electron gained moving upward. By conservation of energy, the photon's energy is necessarily equal to the change in electron energy between the two orbits. Using Einstein's rule, photon energy determines the radiation's frequency. Spectral lines were no longer a mystery.

Despite these advances, several major conceptual problems remained. The realization that ultimately Bohr's model of the atom didn't make sense was perhaps the most glaring of them. According to the laws of physics, electrons in orbits rapidly radiate, spiraling into the nucleus as their energy wanes. Having quantum rules stipulate that electrons must lie in orbits circumvents this difficulty in part, but doesn't explain why all electrons don't jump down to the lowest energy orbit.

The problem was answered in 1925 by the exclusion principle established by the brilliant, precocious, already famous twenty-five-year-old Wolfgang Pauli. At twenty-one, Pauli published a legendary 250-page review article of Einstein's Theory of Relativity. In a 1955 letter to his old friend Max Born, Pauli reminisced about his special relation to Einstein: "I will never forget the speech about me, and for me, that he gave at Princeton in 1945 after I got the Nobel Prize. It was like the abdication of a king, installing me as a kind of elected son, as his successor."

A caustic, relentless critic, Pauli was both admired and loved despite and sometimes for his bluntness. Born in fin de siècle Vienna, Pauli studied in Munich, but spent most of his all too short academic life in the same Zurich university that Einstein had attended during Pauli's first years of life.

Pauli's exclusion principle sets the number of electrons that can be in each atomic orbit and explains why, once that number is reached, any additional electrons have to be in the next higher orbit. For instance, the lowest orbit in an atom has at most two electrons. If only one of the slots is filled, as is the case in hydrogen, there's room for one more. Alternatively, that single one can be stripped away. Hydrogen can easily lose or gain an electron, explaining its chemical activity. Helium on the other hand, with the full complement of two electrons in the lowest orbit, will neither borrow nor lend. Consequently helium is chemically inert, a noble gas. Three-electron lithium has two electrons in the lowest orbit and one in the next orbit up. That outside electron is readily movable, making lithium very reactive. Proceeding from element to element, the exclusion principle describes the level of chemical activity of each type of atom and anticipates what sort of molecular bonds atoms will form.

The principle also explains atoms' stability, why electrons in higher energy orbits don't simply drop to the lowest one. Classical physics says this will happen, but exclusion says the descent can only take place if there is an opening below. The principle's wording is that no two electrons can be in the same quantum configuration—the technical term is *state*. The only reason the lowest orbit even has two states is because electrons have spin, an intrinsic electron property with two labels, up or down. If one of the electrons in the lowest helium orbit is up, the other is down—same orbit, different spin, and hence different state.

The exclusion principle does not determine, however, the electron's energy in its orbit. It doesn't even explain what it means for an electron to move in an orbit. Those problems were solved independently by Werner Heisenberg and Erwin Schrödinger in 1926. The resulting theory is known simply as quantum mechanics, the culmination of twenty-five years of thinking about quanta, waves, and particles. Einstein showed that waves had a dual nature and could be treated as particles, photons. Louis de Broglie, a young Frenchman, suggested in the early 1920s the natural extension: treat electrons as waves. Only certain notes can be played on a plucked string, notes whose wavelength is determined by the length of the string. Similarly, only certain electron waves fit on an atomic orbit. Once the wavelength is known, so is the frequency, and then, by duality, so is the electron's energy. The new quantum mechanics gave a mathematical framework for this kind of thinking and for much more.

The Low-Temperature World

Very low temperature is the quintessential domain of quantum mechanics and still very much a frontier area of research. Though practical applications have not yet been abundant, the physics Nobel Prizes of 1996, 1997, and 2001 were all three awarded for experiments studying the behavior of matter within thousandths of a degree above absolute zero. The richness of the newly uncovered phenomena has much to do with the split between the classical and the quantum physics worlds. Classically, motion comes smoothly to an end as absolute zero is approached. Quantum mechanically, smoothness is impossible. The jump from one quantum configuration to another grows in significance as temperature drops; discontinuities becoming more prominent. While a speedy pebble skips along a bumpy, pocked path, a slow one settles in one hole or another. At high speeds, small differences in hole size or location don't matter; at low speeds they do. As we saw earlier, atomic speed is directly proportional to temperature. Low temperature means low speeds and greater prominence of the quantum world's discontinuities.

There is a further complication in understanding the low-temperature quantum world, one that can be traced back to the uncertainty principle, first formulated in 1927 by twenty-five-year-old Werner Heisenberg. This principle sets an ultimate limit to the simultaneous determination of a particle's speed and its position. Greater accuracy in determining an atom's speed necessarily leads to greater uncertainty in the location of the atom and vice versa.

This principle poses an interesting dilemma in understanding very low temperature. If we carry over from classical physics the notion that absolute zero corresponds to zero atomic speed, we are immediately faced with a conundrum. According to the uncertainty principle, an atom's location is increasingly difficult to specify as its speed diminishes, as temperature approaches zero. Unexpected new phenomena appeared in experiments performed at lower and lower temperatures. To carry the analogy of the pebble a step further, as temperature drops, it becomes increasingly difficult to know with certainty where any one pebble is.

An interesting illustration is the case of liquid helium. In 1910 Kamerlingh Onnes realized that the density of liquid helium reached a maximum at 2.2 degrees Kelvin. He also noticed the boiling liquid becoming strangely quiet. Coming back to the question in the 1920s, Onnes and Leo Dana saw that "near the maximum density something happens to the helium which within a small temperature range is even discontinuous." The bubbling seemed to suddenly stop, the liquid's surface to become smoother. They didn't know what it meant or what was happening.

Their lead was followed in England by Pyotr Kapitsa,

a brilliant Russian and the son of a czarist general. A protégé of Rutherford's, Kapitsa set up a world-renowned low-temperature laboratory in Cambridge, but his work in England came to an end in 1935 when he was detained in the Soviet Union after a visit home. Rumor had it that Stalin wanted him to help in Russia's electrification. When it became clear that Kapitsa was not going to be allowed to leave Russia, Rutherford sent him Cambridge laboratory equipment so he could continue his work.

Meanwhile Jack Allen, a young Canadian physicist who worked on liquid helium problems at the University of Toronto, arrived in Cambridge, hoping to collaborate with Kapitsa. When Kapitsa didn't return to the West, Allen set off on his own with Donald Misener, another young Canadian. In early 1938 the two of them in Cambridge and simultaneously Kapitsa in Moscow announced a discovery just as surprising as superconductivity. They had observed and studied what Onnes and Dana had glimpsed, anomalous behavior of liquid helium below 2.2 degrees. Spinning in a container, the low-temperature helium continues to gyrate without slowing down unless the temperature is raised. Once set in motion, no part can break off. The liquid has become an effective superfluid with no viscosity. The location of a single atom has lost its meaning; all the atoms have become a single "superatom."

Something like the curious behavior displayed by superfluid helium had been envisioned more than a decade before the Allen, Misener, and Kapitsa experiments. Inspired by a 1924 paper from the relatively unknown Bengali physicist Satyendra Bose, Einstein had realized that quantum theory dictates interesting new behaviors among indistinguishable particles, behaviors that are especially prominent at low temperatures. Rather than observing liquid helium, Einstein considered a collection of atoms at low pressure and low temperature, in gas form rather than liquid form. This is conceptually simpler because the interactions between atoms are weaker in the more dilute state of matter. Though this aspect of the experiment is simpler in principle, there are formidable technical problems involved in achieving the kind of transition Einstein discussed. The experiment turned out to be so hard that the necessary techniques were not fully developed until 1995, seventy years after Einstein's conjecture. At that time a group in Boulder, Colorado, finally succeeded in changing 2000 rubidium atoms into a single "superatom" for ten seconds. It created what has come to be known as a Bose-Einstein condensate. At about the same time a similar condensate was made at Rice University in Texas in a lithium gas, and in a sodium gas at MIT.

The path had been long, arduous, and marked by unanticipated difficulties. In 1895 helium was first found on Earth and the lowest temperature reached in a laboratory experiment was more than 10 degrees above absolute zero. By 1995, a hundred years later, when the Bose-Einstein condensate was finally observed, the temperature in the experiment was less than two hundred billionths of a degree above absolute zero.

There is a further complication that Bose and Einstein had not envisioned. Less than a year after the appearance of their papers, Pauli proposed the "exclusion principle." This asserts that no two electrons can be in the same quantum state, almost the direct opposite of the suggested low-temperature grouping of helium atoms into a single superatom. It turns out that one of the many subtleties of quantum mechanics is that there are two radically different ways indistinguishable particles may act, resulting in two very different behaviors near absolute zero. Helium atoms act one way, electrons another.

In quantum mechanics even the subtleties have subtleties. Helium atoms form a superfluid at 2.2 degrees above absolute zero, but this turns out to be true only for normal helium-4, the atom with two neutrons in the nucleus. The rare isotope, helium-3, though chemically indistinguishable from helium-4, obeys the exclusion principle.

Understanding the foundations and implications of quantum mechanics has been one of the twentieth century's great intellectual adventures. The wunderkinder who helped pioneer this trail grew up, became even more famous, and received Nobel Prizes before they were forty. Marvel upon marvel fell on scientists in the wake of these discoveries; problems that seemed insoluble were solved. It was a golden era for those who embraced the new rules of quantum mechanics and a difficult time for those who didn't. There have been other revolutions in twentieth-century science, but perhaps none in which the very underpinnings had to be changed so radically and so quickly. Unfortunately, the shift was incomprehensible or unpalatable to many of the older pioneers.

Einstein's Refrigerator

The 1924 paper on Bose-Einstein condensation was the last truly influential paper Einstein wrote. He was forty-five years old. A new generation was taking over, a generation that built quantum mechanics in large measure on foundations he had helped to lay. Research on quantum science was proceeding at breakneck speed after 1925, but Einstein never fully accepted the turn of events. In a famous December 1926 letter to his friend Max Born, the senior theorist at Göttingen and one of the shapers of the new subject, Einstein wrote, "Quantum mechanics is very impressive. But an inner voice tells me that it is not yet the real thing. The theory produces a good deal but hardly brings us closer to the secret of the Old One. I am at all events convinced that *He* does not play dice."

Einstein went to his death convinced that, even though the theory worked and all its predictions agreed with experiment, there must be some different underlying framework. In expressing this point of view, Einstein carried on legendary debates with many scientists, his old friend Bohr in particular. The distinguished historian of physics Martin Klein puts it this way: "Einstein's colleagues could only regret that he had chosen to follow a path separate from the rest. As Born wrote, 'Many of us regard this as a tragedy—for him, as he gropes his way in loneliness, and for us, who miss our leader and standard-bearer.' "

The year 1925 marks the beginning of quantum mechanics and the end of Einstein's influence as a major force in guiding research. From then until his death in 1955, he pursued the dream of unifying his beloved general relativity and electromagnetism into one all-encompassing theory. As younger theorists went on to explore the ramifications of the new quantum mechanics, Einstein had little intellectual influence on the choices they were making. Clinging to his dream, he was revered but ignored by the young.

In the late 1920s, Einstein was doing more than criticizing the new quantum mechanics and trying to find the unified field theory. He was also building a better household refrigerator. Bizarre as it may seem, this is how the world's greatest scientist was spending at least part of his time. The man who geometrized space and time, the icon of humanity, and the legendary prototype of the absentminded professor was intrigued by a more mundane aspect of temperature, the refrigerator.

Einstein had often felt more at ease as an outsider, intellectually as well as socially. Renouncing his German citizenship in 1896 when he was not yet seventeen, he applied in 1899 for Swiss citizenship and received it in 1901. Even after he moved back to Germany in 1913 to become a professor in Berlin, he retained his Swiss citizenship. As Einstein is said to have quipped, "If relativity is right, the Swiss will say I'm Swiss and the Germans will say I'm German. If relativity is wrong, the Swiss will say I'm German and the Germans will say I'm a Jew."

Shortly after graduation from university in Zurich, unable to obtain an academic position, Einstein was hired as Technical Expert (Third Class) in the Bern Swiss Patent Office. He stayed at the Patent Office until 1909, by which time he was already famous. Apparently Einstein found his position quite interesting: he examined proposals for patents, determined if the inventions, often presented in rough form, functioned properly, and decided if they warranted legal protection. In a nutshell, Einstein's job was assessing the inventions.

In his spare time Einstein discovered the Special Theory of Relativity, the nature of the quantum, and explained Brownian motion. However, the patent office work was sufficiently interesting for Einstein to retain a lifelong fascination with practical devices, how they are assembled, and how they can be improved. A rare combination, Einstein was creative even in his mundane activities.

One day in 1925 Einstein read a story in the newspaper about a family that had been killed by poisonous gases leaking from the pump of their refrigerator. At the time commercial refrigerators were still relatively new, the sort of proposal Einstein would have considered in his Bern office. The principle on which refrigerators work still hasn't changed much. A mechanical compressor acts on a refrigerant gas, liquefying it by compression. In this step the heat generated by liquefaction is released to the outside. The refrigerant is then allowed to expand and evaporate. Since an expanding gas cools, heat is absorbed, but now from the interior chamber of the refrigerator. Effectively the cycle works to remove heat from the inside of the refrigerator and transfer it to the surroundings. It's like one of those heat engines I discussed in chapter 2.

How could such an innocent-sounding device kill people? The answer is that the three refrigerant gases commonly used, all selected for their cooling efficiency, were methyl chloride, ammonia, and sulfur dioxide. All three are toxic. A crack in any one of the seals of the moving parts could lead to leakage of the gas into the surroundings. If this happened, the result could be deadly.

After reading the sad story in the paper, Einstein turned to Leo Szilard, a young friend of his and a student in Berlin. Szilard, then in his mid-twenties, was a bright, energetic physicist who had come to Berlin from his native Hungary to obtain a graduate degree. Like Einstein, he was interested in problems of thermodynamics and statistical mechanics. The two met frequently. At Einstein's suggestion, they decided to also pursue something practical, a better refrigerator. They set about devising ways to minimize the danger of leakage by eliminating as many moving parts as possible, e.g., building a better pump. The main challenge was to increase safety by removing the moving compressor piston. Their innovation was to replace the piston with an alloy formed from a mixture of liquid sodium and potassium. The liquid was a metallic fluid that could be moved up and down inside a sealed casing by a magnetic field induced by an alternating current. The compression of the refrigerant was due to the liquid metal while the expansion of the refrigerant proceeded as usual.

By early 1926, Einstein and Szilard were busy filing patents. Of course, Einstein knew a lot about patents from his Bern days, so he and Szilard could even dispense with the usual patent attorney. The patents were reasonably successful.

Initial negotiations with the Bamag-Beguin Company fell through but Platen-Munters Refrigerating Systems, a division of Electrolux, bought one of their patent applications for the equivalent of $10,000. Later patent applications were also bought, but none went into production. Despite their best efforts, the Einstein-Szilard pump did not equal conventional refrigerators in cooling efficiency, making it more expensive. The initial refrigerator safety dilemma that motivated Einstein and Szilard was solved in 1930 by the development in America of a nontoxic refrigerant, "Freon." Decades later we realized that Freon belongs to a class of gases called chlorofluorocarbons that endanger the ozone layer protecting us from the Sun's ultraviolet radiation. So we need to rethink the refrigeration problem again, but first back to Einstein.

Einstein made no money from his patents. A few months after Hitler became chancellor, both Einstein and Szilard fled Germany. Both landed in the United States, Szilard via Britain, Einstein joining the newly founded Institute for Advanced Study in Princeton, where he remained until his death. His agreement with Szilard stipulated that they would split profits only if Szilard's income rose above a certain level. Until then Szilard would receive all the income. The patent money more than doubled Szilard's meager income and allowed him to save money with which he both lived and helped other refugees fleeing Germany.

Although Einstein enjoyed being an inventor, he left that part of his life behind when he fled from Germany. Perhaps his age was a factor, but he never again formed associations with inventors who were also friends. Szilard is the best-known such example, but there were others. A noted singer, who was also an acquaintance of Einstein's, complained to him of increasing hearing difficulties. In 1928 Einstein turned to Rudolf Goldschmidt, the director of an industrial research laboratory, to help him develop a new type of hearing aid. The patent for a "Device, especially for sound-reproduction equipment, in which changes of an electric current generate movements of a magnetized body by means of magnetostriction" was issued on January 10, 1934. It gave Einstein's address as "Earlier in Berlin, present residence unknown." He had already fled Germany.

Our refrigerator story has two footnotes, both involving Leo Szilard. Szilard, always a visionary, became a leading physicist, a political force, an author, in the end a biologist. He also filed several patents on his own. Notably in 1934, at the dawn of the nuclear age and a full four years before the uranium nucleus was split, he filed a secret Admiralty patent in London (patents in the United Kingdom could be secret if the benefits were assigned to the government) covering the basic process of a nuclear chain reaction. He did so because of his belief that "if a nuclear chain reaction can be made to work it can be used to set up violent explosions." Later, with Enrico Fermi, he filed a patent covering the essential features of a nuclear reactor. Safety is paramount in nuclear reactors; a failsafe cooling method is absolutely necessary. Szilard, who knew all about nuclear reactors, told his fellow scientists he had just what they needed, the Einstein-Szilard pump.

The second footnote concerns a letter Szilard wrote in August 1939. By then, recognizing the potential for building a nuclear bomb, he felt it was imperative to warn his new country's government of its dangers. He decided to write Roosevelt a letter. Fully aware that he needed allies to make his warning heard, he turned to Albert Einstein; a letter from Szilard might be ignored by the president, but not one from the world's greatest scientist. At the time, Einstein's scientific interests lay in the area of unified field theory and not in nuclear physics. While unaware of the potential for destruction

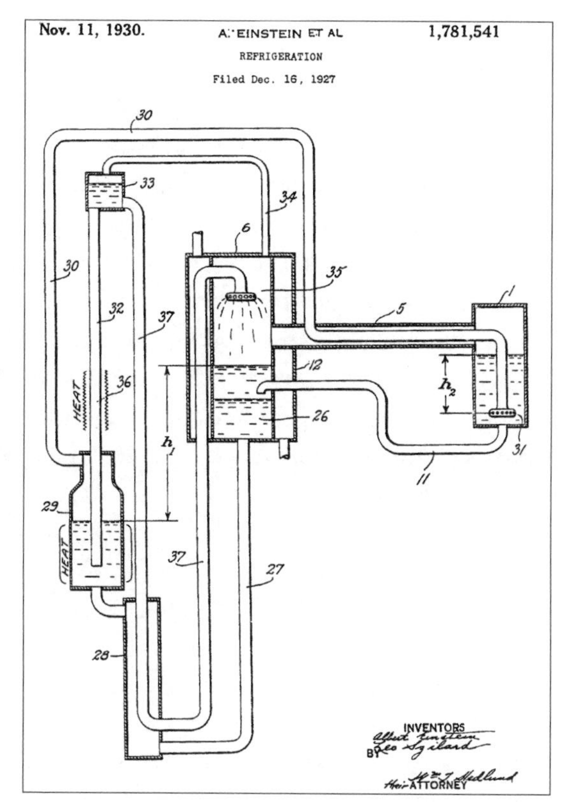

Schematic drawing from the application by Albert Einstein and Leo Szilard for a U.S. patent for their refrigerator

of an atom bomb, Einstein trusted both Szilard's technical knowledge and his political instincts. Szilard drafted the letter and Einstein signed it. This letter is often considered the first step in a chain of events that led to the Manhattan Project and from there to the atom bomb.

Chandra's Journey

In the late 1920s, as the aging Einstein was criticizing the new quantum mechanics (and of course designing refrigerators), an Indian teenager was busy learning the rules of quantum mechanics and relativity. On the afternoon of July 31, 1930, nineteen-year-old Subrahmanyan Chandrasekhar boarded a boat in Bombay destined for Italy, the first stop in Europe on his journey to Cambridge, England. When the boat docked in Venice a little less than three weeks later, Chandra, as he later came to be known, had realized something remarkable about the death of stars like our Sun.

A great deal about the stellar life cycle had already been learned in the previous decade. Scientists agreed on the prediction that the Sun will live for ten billion years, converting hydrogen into helium. After that, it will briefly swell into a cooler larger red giant, engulf the Earth, and then retreat, becoming a white dwarf. With its hydrogen supply exhausted, the Sun will leave behind a core that is a mixture of carbon and oxygen surrounded by helium.

By 1930 the theory of white dwarfs was sufficiently developed to include the effects of quantum mechanics. The star's core temperature, though not high enough to induce further fission, is sufficient to strip all electrons away from their nuclei. Pressed together by gravity's relentless squeeze, these nuclei form a central crystalline-like edifice, a diamond in an ocean of electrons, more precisely what is known as a Fermi sea of electrons.

Prior to 1930, it was believed that all stars ended as white dwarfs. None argued this point more forcefully than Sir Arthur Eddington, the brilliant dean of British astrophysicists and the author of *The Internal Constitution of Stars,* a text Chandra studied in his native Madras. It was one of the three science books he carried with him on his trip to England. During the voyage, Chandra asked himself if the inclusion of relativistic effects would make any difference in the quantum mechanical treatment of white dwarfs. He knew the Pauli exclusion principle made electrons in the Fermi sea resist squeezing by generating an outward pressure, but suspected relativity would not affect the outcome. Much to his surprise, Chandra's calculations indicated that electron pressure with relativity taken into account will not prevent collapse if the star is much bigger than our Sun. A large enough star will not die quietly as a cold white dwarf: it has to follow a different path.

Even today not too many nineteen-year-olds can follow the intricate argument, much less create it. But then there are also not too many three-week boat trips. Chandra made his argument to fellow Cambridge astronomers. It was rejected. In late 1930 he tried to publish a paper on the subject in the *Monthly Notices of the Royal Astronomical Society,* but was unable to. The paper went to the American *Astrophysical Journal.* Chandra worked intermittently on the problem for the next few years, fleshing out every argument, checking every detail. By 1934, now a Ph.D. and a fellow of Trinity College, Cambridge, he could say with conviction that only stars smaller than 1.4 solar masses, now known as the Chandrasekhar limit, end their lives as white dwarfs. Bigger stars continue to collapse.

He felt sure that Eddington, the British champion of Einstein's General Theory of Relativity, would appreciate his reasoning. Eddington had led the 1919 solar eclipse expedition to South America that verified Einstein's great prediction of the bending of starlight by the Sun. In his 1920 presidential speech to the British Society for the Advancement of Science, Eddington pointed out to his fellow astronomers the importance of Einstein's relation between mass and energy in explaining that hydrogen fusion was likely to be the Sun's source of energy. If anybody in Britain could speak about the effects of relativity on stars, it was Eddington.

Sadly, Eddington, who had so often been in the vanguard and whom Chandra thought of as a mentor, simply refused to believe his argument. He insisted that all stars ended as white dwarfs. He mocked Chandra's 1934 proof without understanding it. Though the two scientists remained friends, Eddington's opposition to Chandra's views about star collapse continued unabated until Eddington's death in 1944, delaying but not stopping their acceptance and eventual acclaim by the astronomy community. As Chandra later remembered, "The moral is that a certain modesty toward science always pays in the end. These people (Eddington . . .) terribly clever, of great intellectual ability, terribly perceptive in many ways, lost out because they did not have the modesty to say 'I am going to learn what physics teaches me.' They wanted to dictate how physics should be."

In spite of his early difficulties, Chandra went on to a brilliant career that included writing books about stellar structure and about the collapse of stars. His work continued till the end to have a profound influence on all who read it or heard him speak. Yet, despite receiving all possible honors including the Nobel Prize, he is still most famous for the "Chandrasekhar limit."

Why wasn't the importance of Chandra's argument accepted earlier? There are many reasons. Collapsing stars seemed too remote an object for most physicists; they felt they had more than enough on their plate just understanding the implications of the new quantum mechanics and the details of the nucleus. Also, astronomers were suspicious of the new insights into the structure of matter afforded by quantum mechanics and relativity. The notion that the combination of the two would limit the size of white dwarfs seemed too remote to them. Science was moving forward so quickly that it was hard to assimilate new ideas, to follow the new proposals to their logical conclusion, especially if established scientists didn't support them.

"A certain modesty toward science" is often hard to come by. Stories like Chandra's were true yesterday and will be true tomorrow. The accelerating pace of scientific research doesn't make it any easier. I remember being stunned to learn that Bohr had never met Einstein until April 1920. Berlin and Copenhagen are so close, and the two were already such towering figures. Of course World War I created a barrier, but I still assumed they would have met many times before then. Thankfully the ease of travel and communication today facilitates all kinds of interaction. Nowadays figures like Einstein and Bohr, were they to reappear, would quickly find each other and exchange frequent e-mails. But we would not have the souvenirs that the lost art of letter writing gives us of their first meeting. After it, Einstein wrote,

> Not often in life has a person, by his mere presence, given me such joy as you did. I know why Ehrenfest loves you so. I am now studying your great papers and in doing so—especially when I get stuck somewhere—I have the pleasure of seeing your youthful face before me, smiling and explaining.

Bohr replied:

> For me it is one of the great experiences of life that I can be near you and talk to you, and I cannot say how grateful I am for all the friendliness with which you showed to me on my visit to Berlin. You do not know how great a stimulus it was for me to have the long awaited opportunity of hearing from you personally your views on the very question with which I myself have been busy. I will never forget our con-

versation on the way from Dahlem [a Berlin suburb] to your house.

Six years later, their legendary disagreements on quantum mechanics began, but their love and respect for one another never wavered. These letters remind us that research in science

Albert Einstein and Niels Bohr soon after their first meeting, in 1920

is still a human occupation, as full of pain and joy as any other activity. Winning matters, getting there first can make or break a career; companionship helps in dealing with the pleasures and disappointments. Style matters, taste is important, technique is paramount, creativity is key. All too often these days, government funding also makes the difference between what can and what cannot be done. But sharing also matters, communicating with others, knowing you are not alone.

Ultimately one is confronting nature, looking for her secrets. I confess to an increased feeling of awe as I learn more about science and its endless riches, an awe if anything increased by familiarity. Sometimes that feeling grows slowly in me. Other times it strikes suddenly, never more so than when two widely different disciplines suddenly come to bear on the same problem or when a well-known puzzle is unexpectedly seen in a new light.

I can only imagine the thrill Watson and Crick felt as they realized the implications of the double helix structure of DNA. In their own words, "It has not escaped our notice that the specific pairing we have postulated immediately suggests a possible copying mechanism for the genetic material." When Einstein recognized that the General Theory of Relativity explained the peculiarities in Mercury's orbit, he had heart palpitations. His distinguished biographer Abraham Pais, who knew him well, writes, "This discovery was, I believe, by far the strongest emotional experience in Einstein's scientific life, perhaps in all his life. Nature had spoken to him." There are countless examples: Wegener realizing the congruity of Africa and South America, Galileo measuring the period of the pendulum in Pisa's cathedral. Several, related to temperature, have appeared in this book: Onnes observing mercury turn superconducting at 4 degrees, Rutherford seeing the alpha particles bounce back from the target, Penzias and Wilson detecting 3-degree radiation from outer space, and Corliss observing ten-foot worms on the hydrothermal vents.

In cases like these, the thrill of detection is combined with the eerie pleasure of knowing you are the first person to uncover a truth. More than one scientist, both famous and ordinary, has narrated the strange thrill associated with the fleeting awareness of having discovered a secret of nature. The excitement is great, but it is only one of many in our lives. I'm sure none of us recalls our first steps as an infant, except in the tales of our parents, but that was surely one of our greatest advances. I can still relive in memory the joy of first learning to read, of being able to ride a bicycle. But I also remember the thrill that came on first seeing Cauchy's Theorem in the theory of complex variables and Dirac's equation for the relativistic electron, or realizing there was an absolute zero of temperature.

Into the Future

As I grow older, I feel more the need to integrate the experiences of my life, to find patterns of connection in them that tie me to family, to community. In somewhat the same manner, I find myself looking for links in science, ties of one area to another that remind me of the unity of experience. This is one of the reasons I have enjoyed thinking about temperature. The radioactive alpha particle decay of a few kinds of nuclei heats the Earth's crust, which convectively creates fire just below the surface. Forced up through the ocean bottom in hydrothermal vents, this activity somehow sustains life that evolves to create creatures that think about radioactivity.

One needs to combine the sense of awe with resolve. From our belief in our centrality in the universe we have moved to the realization that we are only specks on a continuum. Surely understanding our place in that universe is one of our noblest pursuits. Perhaps we are only special in realizing how ordinary we truly are. The neutrino is the most insignificant of all elementary particles, the weakest in its interactions, but neutrinos in the aggregate may constitute the universe's dominant mass, greater than all the stars. We have created elaborate devices to raise temperature and are justly proud of them. The tiny "bombardier beetle" oxidizes hydroquinone chemicals in a reaction chamber embedded in its body and then shoots them out at over 200 degrees Fahrenheit through a microscopic nozzle in its posterior. Isn't that a small miracle as well?

Our insignificance doesn't mean we cannot or should not act. The Earth has gone through many temperature ups and downs over billions of years and will go through many more, with or without us. We owe it to ourselves and to the cosmos to be respectful of one another and of the limited powers we have to live in harmony with that greater world.

Scientific research, perhaps not all of it but most of it, continues to be a great pursuit, one of the ways we can, I hope, better the lot of ourselves and other living creatures. More than that, it is an outlet for our dreams, a chance to see more of the connections that nature employs to create the world we live in. There is a famous line in Canto XXVI of Dante's *Inferno* where Ulysses, after having safely landed in Ithaca, calls back his old crew and urges them to join him in setting sail again:

> *Considerate la vostra semenza*
> *Fatti non foste a vivere come bruti,*
> *Ma per seguir virtute e canoscenza.*
>
> (Consider how your souls were sown
> You were not made to live like brutes or beasts,
> But to pursue virtue and knowledge.)

This has to be done with courage, the audacity to set sail into the unknown while preserving, as Chandra warned us, the "modesty" to read the data, to know what the temperatures were, and to learn from them.

I don't know where these voyages will take us. Every generation is inclined to say that it has finally solved the great problems, and that succeeding ones will only mop up. If we are fortunate and wise enough to go on as a species for many millennia, I am tempted to think the twentieth century will be remembered as something special in science, the century in which many of the mysteries of Earth, life, and the cosmos were understood for the first time.

In our laboratories, we have reached temperatures of billions of degrees and we are only billionths of a degree away from absolute zero. Is there more? The only thing in this book that I am absolutely positive about is the answer to that question. It is yes.

References

Introduction: The Ruler, the Clock, and the Thermometer

The quote about Villa Adriana is from Marguerite Yourcenar's *Memoirs of Hadrian* (Farrar, Straus & Giroux, 1954).

The quote by my uncle Emilio is from his autobiography—Emilio Segrè, *A Mind Always in Motion* (University of California Press, 1993). A family memoir written by Emilio's son Claudio may also be of interest. The book is Claudio Segrè's *Atoms, Bombs and Eskimo Kisses* (Viking, 1995).

Chapter 1: 98.6

For the workings of the hypothalamus, I consulted chapter 37, "The Hypothalamus: An Overview of Regulatory Systems," by J. P. Card et al. in *Fundamental Neuroscience,* edited by Michael Zigmond et al. (Academic Press, 1999). That's where the Cushing quote is from.

The source for the brief history of fans is my oft-consulted 11th edition of the *Encyclopaedia Britannica.*

The bee-fanning description is in E. O. Wilson's *Sociobiology, The New Synthesis* (Harvard University Press, Belknap Press, 1975).

The story about Eddy Merckx is from a column by Philip Morrison entitled "Air-Cooled" in *Scientific American,* October 1997, p. 149.

The description of the Sahara can be traced back to W. Langewiesche's *Sahara Unveiled* (Vintage, 1991).

Carl Gisolfi and Francisco Mora's very interesting book, *The Hot Brain* (MIT Press, 2000), is an excellent source of information about temperature regulation in humans and other animals. That's where I learned about the Hong study of the *ama* divers.

The quotes from Benjamin Franklin's writings are from *Benjamin Franklin—New World Physicist,* by Raymond Seeger, in the Pergamon Press series *Men in Physics.* This volume was published in 1973.

The story about the swimmer Lynne Cox is from the August 23, 1999, issue of the *New Yorker,* by C. Sprawson.

Knut Schmidt-Nielsen's *Animal Physiology,* 5th edition (Cambridge University Press, 1997), has a wide-ranging discussion of thermal effects in animals, including most of the information I present on the subjects of harp seals and emperor penguins. Schmidt-Nielsen has also written *Desert Animals: Physiological Problems of Heat and Water* (Oxford University Press, 1964; reprinted by Dover Publications, 1979).

Apsley Cherry-Garrard's book is entitled *The Worst Journey in the World* and was reissued by Carroll and Graf in 1997.

The article about the sauna at the South Pole was by Will Silva in the November 1999 *Harvard Magazine.*

A standard modern source on fever and medicine is the section on "Alterations in Body Temperature," by Jeffrey Gelfand and Charles Dinarello, in the 14th edition of *Harrison's Principles of Internal Medicine,* edited by Anthony Fauci et al. (McGraw-Hill, 1998).

An excellent description of Galenic medicine and much more is contained in Daniel J. Boorstin's delightful *The Discoverers* (Vintage, 1985).

The article about four saints as illustrations of the four humors, "Dürer's Diagnoses," is by M. Kemp in *Nature* 391, January 22, 1998, p. 341.

The biography of Pasteur from which I quote is by René Dubos and is entitled *Louis Pasteur—Free Lance of Science* (Charles Scribner's Sons, 1960).

The article in the January 25, 2001, issue of *Nature* about the genome sequence of *E. coli* O157:H7 is by Nicole Perna et al., p. 529.

A discussion of Wagner-Jauregg's contribution is contained in J. Licino, "What Makes One Tic?," *Science* 286, October 1, 1989,

p. 56, and the quote about treating mental patients with fever is taken from an interview with Robert Garber, former president of the American Psychiatric Association. It is quoted by Sylvia Nasar in her biography of the mathematician John Nash: *A Beautiful Mind* (Simon and Schuster, 1998).

P. A. Mackowiak's ideas about fever are proposed in an article of his entitled "Fever: Blessing or Curse: A Unifying Hypothesis," published in the *Annals of Internal Medicine* 120, 1994, p. 1037.

Some articles on the evolution of the hypothalamus are: C. Zimmer, "In Search of Vertebrate Origins: Beyond Brain and Bone," *Science* 287, March 3, 2000, p. 1579; L. Z. Holland and N. D. Holland, "Chordate Origins of the Vertebrate Central Nervous System," *Current Opinion in Neurology* 9, 1999, p. 596; T. C. Lacalli and S. J. Kelly, "The Infundibular Balance Organs in Amphioxus larvae and Related Aspects of Cerebral Vesiole Organization," *Acta Zoologica* 81, 1999, p. 37.

For some general references on drosophila, see Robert E. Kohler, *Lord of the Fly: Drosophila Genetics and the Experimental Life* (University of Chicago Press, 1994); Jonathan Weiner, *Time, Love, Memory* (Vintage, 1999); G. M. Rubin and E. B. Lewis, "A Brief History of Drosophila's Contribution to Genome Research," *Science* 287, March 24, 2000, p. 2216.

For references on heat shock proteins, see W. J. Welch, "How Cells Respond to Stress," *Scientific American,* May 1993, p. 56; J. R. Ellis and S. Van der Vries, "Molecular Chaperones," *Annual Review of Biochemistry* 60, 1991, p. 321; P. C. Turner et al., *Molecular Biology* (BIOS, 1997) p. 188; R. Morimoto, A. Tissieres, and G. Georgeopoulos, *Stress Proteins in Biology and Medicine* (Cold Spring Harbor Press, 1990).

For the central dogma of molecular biology, see Francis Crick's *What Mad Pursuit* (Basic Books, 1988). For the communality of genes, see François Jacob's *Of Flies, Mice and Men* (Harvard University Press, 1998).

Chapter 2: Measure for Measure

The metric equivalents of the Sumerian *zir,* the Akkadian *ubanu,* the Assyrian *imeru,* and the Jewish *gomor* are given in an article by my father, Angelo Segrè, "Babylonian, Assyrian and Persian Measures," in the *Journal of the American Oriental Society* 64, 1944, p. 73.

For a discussion of the search for early human fires, see B. Wuethrich, "Geological Analysis Damps Ancient Chinese Fires," *Science* 281, July 10, 1998, p. 165. For one about the evidence of early cooking of tubers, see E. Pennisi, "Did Cooked Tubers Spur the Evolution of Big Brains?," *Science* 283, March 26, 1999, p. 2004.

For a general overview of the development of civilization, there is no better book I know than Jared Diamond's wonderfully enlightening *Guns, Germs, and Steel* (W. W. Norton, 1997).

For an interesting discussion of iron smelting in sub-Saharan Africa that first started at about the same time it did in the Fertile Crescent and that persisted relatively unchanged until the twentieth century, see F. Van Noten and R. Raymaekers, "Early Iron Smelting in Central Africa," *Scientific American,* June 1988, p. 64.

For a discussion of the history, physics, and chemistry of pottery and glazes, see Pamela Vandiver's article "Ancient Glazes" in *Scientific American,* April 1990, p. 20.

For references on Galileo, see Stillman Drake's *Galileo* (Oxford University Press, 1996) and *Galileo at Work: His Scientific Biography* (University of Chicago Press, 1978). For a recent bestseller about Galileo, see Dava Sobel's very readable *Galileo's Daughter* (Walker, 1999).

For a discussion of the impact of Santorio and for the quote from Hooke, see Daniel J. Boorstin's *The Discoverers* (Vintage, 1985).

W. Knowles Middleton's *A History of the Thermometer and Its Use in Meteorology* (Johns Hopkins Press, 1966) is the definitive book on the history of the thermometer. The quotes from Viviani and Newton are both drawn from this book.

The descriptions of Count Rumford, including the quote from the letter to Pictet, are from Sanborn Brown's *Benjamin Thompson—Count Rumford on the Nature of Heat* (Pergamon Press, 1967).

An excellent modern treatment of both the scientific and historical questions regarding the nature of heat is contained in S. G. Brush, *The Kind of Motion We Call Heat,* 2 vols. (North-Holland, 1976).

For a biography of Dalton, see *John Dalton and the Atom* by Frank Greenaway (Cornell University Press, 1966).

Disraeli's quote is from Freeman Dyson's illuminating *Infinite in All Directions* (Harper & Row, 1988).

Carnot's quote is from S. Carnot, *Reflections on the Motive Power of Fire*, translated by E. Medoza (Dover Publications, 1960).

I have taken some small liberties in the interest of readability. For instance, Maxwell would not have referred to the moving constituents of a gas as molecules. The concept became clear only later, but the essential argument is unchanged.

Emilio Segrè's *From Falling Bodies to Radio Waves* (W. H. Freeman, 1984) is an excellent introduction to the history of physics, containing as well references back to the works of Carnot, Kelvin, Mayer, Helmholtz, Clausius, Boltzmann, and Gibbs. I also enjoyed the discussion in Nathan Spielberg and Bryon Anderson's *Seven Ideas That Shook the Universe* (John Wiley, 1987).

The quote by François Jacob is from page 194 of a book I will refer to again in chapter 4. It is entitled *The Logic of Life* (Pantheon Books, 1973).

Chapter 3: Reading the Earth

The story of the icebreaker *Yamal* is told in the *New York Times*, August 19 and August 29, 2000. The story of the *St. Roch* appeared in the *Times* on September 5, 2000, and the story of the polar bears on November 12, 2000.

To read about the West Antarctic Ice Sheet (WAIS), see Michael Oppenheimer's "Global Warming and the Stability of the West Antarctic Ice Sheet," *Nature* 393, May 28, 1998, p. 325.

The classic book describing the shift entailed by Copernicus's view of the universe is Thomas Kuhn's *The Copernican Revolution* (Harvard University Press, 1957).

The article about the alignment of the pyramids is by Kate Spence, "Ancient Eygptian Chronology and the Astronomical Orientation of Pyramids," in *Nature* 408, November 16, 2000, p. 320, with an accompanying commentary by Owen Gingerich on p. 297 of the same issue.

Some of the material leading up to the theories of ice ages and climatic changes is covered in Daniel J. Boorstin's *The Discoverers* (Vintage, 1985), already cited in the previous chapter, and also in Timothy Ferris's *Coming of Age in the Milky Way* (William Morrow, 1988).

The story of the ice ages, ranging from early astronomical conjectures through the work of Agassiz, Croll, and Milankovitch up to the present, is very well told by John Imbrie, a distinguished geophysicist-oceanographer, and science writer Katherine Palmer Imbrie in *Ice Ages—Solving the Mystery* (Harvard University Press, 1979). Part of the story—pre-1863—has also been told recently by Edmund Bolles in *Ice-Finders: How a Poet, a Professor and a Politician Discovered the Ice Age* (Counterpoint Press, 2000). The trio in that book is Kane, an early Arctic explorer; Agassiz; and Lyell. I'm not sure Lyell would have liked being called a politician, but then he might not have liked being called a lawyer either. He was, of course, a great geologist. Some further reading of interest is Loren Eiseley's *Darwin's Century* (Anchor Books, 1961), and Charles Darwin's *The Autobiography of Charles Darwin* (W. W. Norton, 1958).

Some articles relating to Milankovitch cycles that might be of interest are J. Hays, J. Imbrie, and N. Shackleton, "Variations in the Earth's Orbit: Pacemaker of the Ice Ages," *Science* 194, December 10, 1976, p. 1121; R. Muller and G. MacDonald, "Glacial Cycles and Astronomical Forcing," *Science* 277, July 11, 1997, p. 215; S. Kortenkamp and S. Dermott, "A 100,000 Year Periodicity in the Accretion Rate of Interplanetary Dust," *Science* 280, May 8, 1998, p. 874.

As far as climate change is concerned, it's clear from the text that I have been influenced and impressed by Wallace Broecker's writings. When you study a field, one author often catches your eye. I think the choice was a wise one in this case. I have enjoyed reading W. Broecker, *How to Build a Habitable Climate* (Eldigio Press, 1987), as well as some of his articles, e.g., W. Broecker, "Chaotic Climate," *Scientific American*, November 1995, p. 62; W. Broecker and G. Denton, "What Drives Glacial Cycles?," *Scientific American*, January 1990, p. 46; W. Broecker, "Thermohaline Circulation, the Achilles' Heel of Our Climate System: Will Man-Made Carbon Dioxide Upset the Current Balance?," *Science* 278, November 28, 1997, p. 1582.

The literature on climate change is vast, but books that may be useful are T. Graedel and P. Crutzen, *Atmosphere, Climate and Change* (W. H. Freeman, 1995); and, on the subject of El Niño, M. Glantz, *Currents of Change: El Niño's Impact on Climate and Society* (Cambridge University Press, 1996). For an older encyclopedic study, see H. H. Lamb's *Climate, Past, Present and Future,* 2 vols. (Methuen, 1972); and for an interesting historical review, see E. Le Roy Ladurie's *Times of Feast, Times of Famine* (Farrar, Straus & Giroux, 1971). In addition, I recommend a set of more specialized articles: R. Alley and M. Bender, "Greenland Ice Cores: Frozen in Time," *Scientific American,* February 1998; Ping Chang and David Battisti, "The Physics of El Niño," *Physics World,* August 1998, p. 41; T. Crowley, "Causes of Climate Change over the Past 1000 Years," *Science* 289, July 14, 2000, p. 270; P. Epstein, "Is Global Warming Harmful to Health?," *Scientific American,* August 2000; T. Karl, N. Nichols, and J. Gregory, "The Coming Climate," *Scientific American,* May 1997, p. 78; R. Kerr, "Warming's Unpleasant Surprise: Shivering in the Greenhouse," *Science* 281, July 10, 1998, p. 156; M. McElroy, "A Warming World," *Harvard Magazine,* December 1997, p. 35; D. Oppo, "Millennial Climate Oscillations," *Science* 278, November 14, 1997, p. 1244; J. Toggweiler, "The Ocean's Overturning Circulation," *Physics Today,* November 1994, p. 45.

A geology textbook would be useful for further study. An excellent one that also contains substantial sections devoted to global warming is L. Kump, J. Kasting, and R. Crane's *The Earth System,* published in 1999 by Prentice-Hall.

Two recent and very readable articles on the subject of the climates of Mars and Venus are M. Bullock and D. Grinspoon, "Global Climate Change on Venus," *Scientific American,* March 1999, p. 50; and T. Karl, N. Nichols, and J. Gregory, "The Coming Climate," *Scientific American,* May 1997, p. 78. For the latest about Mars, see the review by Richard Kerr, "A Wetter, Younger Mars Emerging," *Science* 289, August 4, 2000, p. 714.

"Fire and Ice" is from Robert Frost, *Collected Poems* (Henry Holt, 1939).

The description of Cavendish is from the 11th edition of the *Encyclopaedia Britannica,* as is the Tyndall quote.

For the history of global warming, see Spencer Weart, "The Discovery of the Risk of Global Warming," *Physics Today,* January 1997, p. 34; Paul Crutzen and Veerabhadran Ramanathan, "The Ascent of Atmospheric Sciences," *Science* 290, October 13, 2000, p. 299.

The United Nations report, "Climate Change 2001: The Scientific Basis," can be found on the Internet at http://www.ipcc.ch. For an up-to-date discussion of this report and the one prepared by the National Energy Policy Development Group, written by Dick Cheney et al., see Bill McKibben's article, "Some Like It Hot," in the July 5, 2001, issue of the *New York Review of Books.*

For worst-case scenarios of global warming, see P. M. Cox et al., "Acceleration of Global Warming Due to Carbon-Cycle Feedbacks in a Coupled Climate Model," *Nature* 408, November 9, 2000, p. 185.

For alternatives to carbon dioxide reduction, see the article by A. Revkin in the *New York Times,* August 19, 2000, and J. Hansen's article in *Proceedings of the National Academy of Science* 97, 2000, p. 9895.

For a report on the Hague meeting, see David Dickinson, *Nature* 408, November 30, 2000, p. 503.

For the attitude of developing nations, China and India in particular, see "Equity Is the Key Criterion for Developing Nations," by Ehsan Masood, in *Nature* 390, November 20, 1997, p. 216.

The article I quoted on equity is entitled "Equity and Greenhouse Gas Responsibility" by Paul Baer et al. in *Science* 289, September 29, 2000, p. 2287.

Chapter 4: Life in the Extremes

The poem at the beginning of chapter 4 is from Robert Lowell's "The Quaker Graveyard in Nantucket," in *Lord Weary's Castle* (Harcourt and Brace, 1946).

The quote by Wegener is from an article, "Alfred Wegener and the Hypothesis of Continental Drift," published in *Scientific American,* February 1975, p. 77.

Robert Ballard knows more about diving in *Alvin* than anybody else in the world. He has also written, with Will Hively, a delightful book entitled *The Eternal Darkness: A Personal History of Deep-Sea Exploration,* published in 2000 by the Princeton University Press. This is the source of much of the material in the beginning of this chapter, including the quote about Corliss and Van Andel and the one by Francheteau. Victoria Kaharl has also written a book called *Water Baby: The Story of* Alvin (Oxford University Press, 1990) chronicling *Alvin*'s adventures.

I found a description of the mining activities off Papua New Guinea in an article by William Broad in the "Science Times" section of the *New York Times,* December 30, 1997.

The definitive text on hydrothermal vents, at least as far as any text can be described as definitive in such a complex, interdisciplinary, fast-moving field, is Cindy Lee Van Dover's authoritative *The Ecology of Deep-Sea Hydrothermal Vents,* published by the Princeton University Press in 2000.

For a recent readable article about Pompeii worms by experts, see S. C. Cary, T. Shank, and J. Stein, "Worms Bask in Extreme Temperatures," *Nature* 391, February 5, 1998, p. 545.

The quote describing Pasteur's philosophy is from René Dubos's biography, *Louis Pasteur—Free Lance of Science* (Charles Scribner's Sons, 1960).

For a good review of extremophiles, see the *Scientific American,* April 1997, p. 82, article by Michael Madigan and Barry Marrs entitled "Extremophiles," and the review article "Life in Extreme Environments," by Lynn Rothschild and Rocco Mancinelli, in *Nature* 409, February 22, 2001, p. 1092.

Tommy Gold's initial proposal of a large underground biomass was published in the *Proceedings of the National Academy of Sciences* 89, 1992, p. 6045.

For a discussion of the nuclear winter scenario, see P. R. Ehrlich et al., "Long-Term Biological Consequences of Nuclear War," *Science* 222, December 23, 1983, p. 1293. H. W. Jannasch and M. J. Mottl suggest in *Science* 229, August 16, 1985, p. 717, that in such a case, "the chance of survival of such ecosystems is the highest of any community in the biosphere." Of course, the ecosystems they refer to are those on the hydrothermal vents.

The story of the explosion of Krakatau is told very well by David Quammen in *The Song of the Dodo* (Simon and Schuster, 1997).

For a discussion of the release of methane and its effects on life, see "Did a Blast of Sea-Floor Gas Usher in a New Age?" by Richard Kerr in *Science* 275, February 28, 1997, p. 1267; and "Methane Fever" by Sarah Simpson in *Scientific American,* February 2000, p. 24.

Paul Hoffman and Daniel Schrag cover the subject of "Snowball Earth" in the *Scientific American* article of the same name. It appeared on page 68 of the January 2000 issue. The exchange I allude to between these authors and William Hyde et al. is from *Nature* 409, January 18, 2001, p. 306.

J. B. Corliss, J. A. Baross, and S. E. Hoffman, "Hypothesis Concerning the Relationship Between Submarine Hot Springs and the Origin of Life on Earth," *Oceanologica Acta,* 1981, p. 59.

A good article about Carl Woese is one by Virginia Morell entitled "Microbiology's Scarred Revolutionary," published in *Science* 276, May 2, 1997, p. 699. It includes the quote I give describing the initial adverse reactions to Woese's proposal.

An up-to-date review of Archaea is contained in the book by J. L. Howland entitled *The Surprising Archaea: Discovering Another Domain of Life* (Oxford University Press, 2000).

Carl Woese's quote is from an article he wrote entitled "The Universal Ancestor" that appeared in *Proceedings of the National Academy of Sciences* 95, June 9, 1998, p. 6854.

The letter from Newton to Bentley is reprinted in *Theories of the Universe,* edited by Milton Munitz (Free Press, 1957).

For a recent discussion of the origins of life and the possible exchanges between branches of the tree of life, see evolutionary biologist W. Ford Doolittle's article, "Uprooting the Tree of Life," in the February 2000 issue of *Scientific American.* A good geology text might also be helpful here, since such books also discuss the origins of life. I already referred to *The Earth System* by Kump, Kasting, and Crane. Another one is *The Earth Through Time* by Harold Levin (Saunders College Publishing, 1999). Both discuss plate tectonics. An astronomy text would probably also be helpful. A very good one is William Kauffman and Roger Freedman's *Uni-*

verse (W. H. Freeman, 1998). That also happens to be where I found the Thomas Jefferson quote.

The tale of the meteor-collision-induced extinction of the dinosaurs is very well told in Walter Alvarez's *T. Rex and the Crater of Doom* (Princeton University Press, 1997).

Details about the collision 250 million years ago have two parts: the first describing the discovery of the crater in Western Australia and the second a chemical analysis of the rock. The first, by a team whose leader was Arthur Mory, can be found in William Broad's story in the April 25, 2000, edition of the *New York Times*. The second is covered in Richard Kerr's article, "Whiff of Gas Points to Mass Extinction," in *Science* 291, February 23, 2001, p. 1469.

The evidence of liquid water in the zircon crystals is contained in two recent articles in *Nature*. One is "Evidence from Detrital Zircons for the Existence of Continental Crust and Oceans on the Earth 4.4 Gyr. Ago," by S. Wilde et al., *Nature* 409, January 11, 2001, p. 175; and the other is "Oxygen-Isotope Evidence from Ancient Zircons for Liquid Water at the Earth's Surface 4,300 Myr. Ago," by S. Mojzsis et al., *Nature* 409, January 11, 2001, p. 178. Both also make the point that a crust had already formed at that time, i.e., the Earth's surface was not molten rock.

An excellent review article, with abundant references, on the subject of early life is E. G. Nisbet and N. H. Sleep's "The Habitat and Nature of Early Life" in *Nature* 409, February 22, 2001, p. 1083.

Evolution of Hydrothermal Ecosystems on Earth (*and Mars?*), edited by G. Bock and J. Goode (John Wiley, 1997), is the report of a symposium held in January 1996 that touches on several points raised in this whole chapter.

The evidence for survival of amino acids in space travel is presented in "Can Amino Acids Beat the Heat?" by R. Irion, in *Science* 288, April 28, 2000, p. 165.

A good description of ALH84001, as well as other topics relating to extraterrestrial life, is contained in Bruce Jakosky's *The Search for Life on Other Planets* (Cambridge University Press, 1998).

For a discussion of panspermia, see *Life Itself, Its Origins and Nature,* by Francis Crick, published by Simon and Schuster in 1981. The book is by now twenty years old, but remains very readable. More recent references can be traced from the Nisbet and Sleep review article cited earlier above.

Donald Brownlee and Peter Ward, in *Rare Earth* (Springer-Verlag, 2000) make the argument that Earth may be the only possible place that life could have evolved, but the caveat is that by "life," they mean complex life. A less controversial discussion of the origins of life covering similar material is in David Koerner and Simon Le Vay's *Here Be Dragons* (Oxford University Press, 2000).

The possibility of life on the Jovian moons is explored in R. Pappalardo, J. Head, and R. Greely, "The Hidden Oceans of Europa," *Scientific American,* October 1999, p. 54; and in T. Johnson, "The Galileo Mission to Jupiter and Its Moons," *Scientific American,* February 2000, p. 40. The recent evidence for water on Ganymede is discussed by P. Schenk et al. in "Flooding of Ganymede's Bright Terrains by Low-Viscosity Water-Ice Lavas," *Nature* 410, March 1, 2001, p. 57. G. Vogel, "Expanding the Habitable Zone," *Science,* October 1, 1999, p. 70, is a good introduction to an extensive special section on "Planetary Systems" within that issue of *Science* magazine.

For recent accounts of the research on Lake Vostok, see Warwick Vincent, "Icy Life on a Hidden Lake," *Science* 286, December 1, 1999, p. 2094; as well as Frank Carsey and Joan Horvath, "The Lake That Time Forgot," *Scientific American,* October 1999, p. 62; and Richard Stone, "Lake Vostok Probe Faces Delays," *Science* 286, October 1, 1999, p. 36. I also used as a reference a story by Robert Hotz that appeared in the March 4, 2001, *Los Angeles Times.*

Chapter 5: Messages from the Sun

I recommend that bibliophiles looking for some of Gamow's books intended for the general public try *The Planet Called Earth* (Viking, 1963), *One Two Three . . . Infinity* (Viking, 1961), or his last book, *A Star Called the Sun,* published by Viking in 1964. The books are, of course, dated, but they give you a feeling for the progress in the field. He also wrote a series of books about a mythical Mr. Tompkins. Finally, there is Gamow's own autobiography, *My World Line: An Informal Autobiography* (Viking, 1970).

For a general reference on astronomy, I recommend William

Kaufmann and Roger Freedman's *Universe* (W. H. Freeman, 1999), and Jay Pasachoff's *Astronomy: From the Earth to the Universe* (Saunders, 1991). For one on cosmology, see John Hawley and Katherine Holcomb's *Foundations of Modern Cosmology* (Oxford University Press, 1998). Topics on thermodynamics can be pursued in Richard Feynman's fabulous, but not always easy, *Lectures on Physics* (Addison-Wesley, 1963).

The quote from Galileo is in Stillman Drake's translation of Galileo's first letter on sunspots (cf. S. Drake, *Discoveries and Opinions* [Anchor, 1957]). It is also quoted in the recent book on Galileo by Dava Sobel, *Galileo's Daughter* (Walker, 1999).

The quotes from Rutherford are in his biography by A. S. Eve, *Rutherford* (Cambridge University Press, 1939).

The quote by Hans Bethe is from a 1996 talk entitled "Influence of Gamow on Early Astrophysics and on Early Accelerators in Nuclear Physics" delivered at the George Gamow Symposium and published by Bookcrafters' Astronomical Society of the Pacific Conference Series in 1997.

John Updike's poem was published in *Telephone Poles and Other Poems* (Alfred A. Knopf, 1963). It originally appeared in the *New Yorker.*

For a history of the solar neutrino problem, see J. Bahcall's *Neutrino Astrophysics* (Cambridge University Press, 1989) or a recent essay by him, "How the Sun Shines," published in June 2000, available at http://www.nobel.se/. For a very recent account of the observation of solar neutrinos, including a description of the results obtained by the Sudbury Neutrino Observatory, see John Bahcall's article "Neutrinos Reveal Split Personalities" in *Nature* 412, July 5, 2001, p. 29.

Tycho Brahe's quote is from D. Clark and F. Stephenson, *The Historical Supernovae* (Pergamon Press, 1977) and is also quoted in Timothy Ferris's book cited earlier. I have relied for physics history on Emilio Segrè's *From Falling Bodies to Radio Waves* (University of California, 1984).

Yang Wei-T'e's quote is given in the book *Universe* by Kaufmann and Freedman cited earlier.

The faint sun problem is discussed in L. Kump, J. Kasting, and R. Crane's *The Earth System*, published by Prentice-Hall in 1999.

The Korycansky solution of having an asteroid nudge the Earth outward is reported in *Scientific American,* June 2001, p. 24.

For an account of the detection of the neutrinos from a supernova written by one of the key participants, see A. K. Mann's *Shadow of a Distant Star* (W. W. Norton, 1997).

For the recent announcement of the discovery of magnetars, see S. R. Kulkarni and C. Thompson, "A Star Powered by Magnetism," *Nature* 393, May 21, 1998, p. 215.

The definitive popular book on black holes is Kip Thorne's *Black Holes and Time Warps—Einstein's Outrageous Legacy* (W. W. Norton, 1994). It's a very readable book by one of the experts in the field. A very readable discussion of superstrings is Brian Greene's recent best-seller *The Elegant Universe: Superstrings, Hidden Dimensions and the Quest for the Ultimate Theory* (W. W. Norton, 1999).

There are many popular or semipopular books about the early universe. Among these are the small classic, *The First Three Minutes,* by Steven Weinberg (Basic Books, 1977) and the recent *The Inflationary Universe* by Alan Guth (Addison-Wesley, 1997). Guth is the man who introduced the notion of the inflationary universe, but the book also has a great deal of information about related topics such as the Cosmic Microwave Background Radiation (CMBR). This topic is still very much at the heart of cosmology, a subject that is becoming increasingly what scientists call data-driven. The most recent set of data (published in *Nature* on April 27, 2000) comes from "Balloon Observations of Millimetric Extragalactic Radiation and Geophysics," better known as BOOMERANG—a set of microwave detectors mounted on a large helium balloon that circled the South Pole for ten days. The results confirm the general picture of the expansion of the universe and support the notion of inflation, but have puzzling features. This is a field that is going to continue to be very exciting for at least the next decade and is likely to provide further surprises. As an example of recent shifts in thinking, see the December 16, 1998, issue of *Science:* "The Accelerating Universe" was listed by the magazine as the top scientific breakthrough of 1998. The Penzias-Wilson story is also very well told in Jeremy Bernstein's *Three Degrees Above Zero: Bell Labs in the Information Age* (Scribner's, 1984). Good books on the early universe written by experts in the field are Joseph Silk's *The Big Bang* (W. H. Freeman, 1980); E. W. Kolb's *Blind Watchers of the Sky* (Addison-Wesley, 1996); Marcelo Gleiser's *The Dancing Universe* (Dutton, 1997); Martin Rees's *Before the Beginning* (Addison-Wesley, 1997); M. Longair's *Our Evolving Universe* (Cambridge University Press, 1990); Stephen Hawking's *A Brief History of Time* (Bantam Books, 1988); John Barrow's *The Origin of the Universe* (Basic Books, 1994); and Leon Lederman and David Schramm's *From Quarks to the Cosmos* (W. H. Freeman, 1989). The abundance of books, many of them recent, means I don't have to give journal article references, but it might be of interest to check the *Scientific American*'s Special Report on Cosmology from January 1999 and the recent crystal-ball gazing by Martin Rees in "Exploring Our Universe and Others" in a special issue of *Scientific American* (December 1999) devoted to "What Science Will Know in 2050." I tend to know the experts' books better, but there are also several books written by science reporters/essayists. Some particularly good ones are Timothy Ferris's *Coming of Age in the Milky Way* (Anchor, 1988) and his recent *The Whole Shebang: A State-of-the-Universe(s) Report* (Simon and Schuster, 2000); Michael Lemonick's *The Light at the Edge of the Universe* (Villard, 1993); and Dennis Overbye's *Lonely Hearts of the Cosmos* (HarperCollins, 1991). There are also several very good books by John Gribbin, one of *Nature*'s editors; these include *The Case of the Missing Neutrinos* (First Fromm, 1998).

The ekpyrotic universe is described in J. Khoury B. Ovrut, P. Steinhardt, and N. Turok's article "The Ekpyrotic Universe: Colliding Branes and the Origin of the Big Bang," in *Physical Review* D64 (2000), p. 123522.

Chapter 6: The Quantum Leap

The standard text on low-temperature physics written by an expert in the field with a general audience, at least of physicists, in mind is K. Mendelssohn's *The Quest for Absolute Zero: The Meaning of Low-Temperature Physics* (Taylor and Francis, 1977). A recent good book with a popular audience very definitely in mind is T. Schactman's *Absolute Zero and the Conquest of Cold* (Houghton Mifflin, 1999).

Dalton's quote is from the 11th edition of the *Encyclopaedia Britannica.*

J. M. Thomas's *Michael Faraday and the Royal Institution* (Adam Hilger Press, 1991) gives the history of Faraday's relation to the Royal Institution. The quote from Faraday and the quote about Davy are from this book.

The geology and astronomy texts I referred to in earlier chapters continue to be useful here. For specific information on drilling holes in the Earth, see Kevin Krakick, "New Drills Augur a Great Leap Downward," *Science* 283, February 5, 1999, p. 781.

A couple of articles written by people who knew the Leiden Laboratories very well are J. de Nobel, "The Discovery of Superconductivity," *Physics Today,* September 1996, p. 40; and R. de Bruyn Ouboter, "Heike Kamerlingh Onnes's Discovery of Superconductivity," *Scientific American,* March 1997, p. 98.

The story of Kamerlingh Onnes's funeral is quoted from reminiscences by the well-known Dutch physicist H. B. G. Casimir in his *Haphazard Reality: Half a Century of Science* (Harper and Row, 1983).

The quote of Max Planck is from Armin Hermann, *The Genesis of Quantum Theory* (MIT Press, 1971).

There are many very good books dealing with Einstein's life and contributions. From a scientist's point of view, the most impressive is Abraham Pais's magisterial *Subtle Is the Lord* (Oxford University Press, 1982). See also Pais, *Einstein Lived Here* (Oxford University Press, 1994). Pais was himself a professor at the Institute for Advanced Study and knew Einstein well. The quote of the letter to Max Born, the quote of the letter to Niels Bohr, and the Goldschmidt hearing aid story are taken from this book. A recent book I enjoyed a great deal is Dennis Overbye's *Einstein in Love* (Viking, 2000). The story of blackbody radiation and of Einstein's contribution to the development of quantum theory is very well told in this book. It contains detailed references for further reading. I also enjoyed Ronald Clark's illustrated biography *The Life and Times of Albert Einstein* (Harry Abrams, 1984) in part because of the many pictures. Bohr's letter to Einstein is printed in this book.

There are many books that deal with quantum mechanics and I am hard-pressed to recommend only one or two of them. For a book with a historical bent, I can suggest Abraham Pais's *Niels Bohr's Times* (Clarendon Press, 1991). The letter from Pauli to Born is quoted in *Physics Today,* February 2001, p. 43, in an article titled "Wolfgang Pauli," by K. von Meyen and E. Schucking. A short article that succinctly describes the highlights of quantum theory is Daniel Kleppner and Roman Jackiw's "One Hundred Years of Quantum Physics," *Science* 289, August 11, 2000, p. 893. This is part of *Science*'s special Pathways of Discovery series. Suggested further readings are given there, including biographies of Dirac, Heisenberg, and Schrödinger.

The quote attributed to Martin Klein is from M. Klein, "Einstein and the Development of Quantum Physics" in *Einstein: A Centenary Volume,* edited by A. P. French (Harvard University Press, 1979).

The story of the Einstein-Szilard refrigerator is told in E. Dannen, "The Einstein-Szilard Refrigerators," *Scientific American,* January 1997, p. 90.

Chandra's story is told superbly by a physicist who knew him well and was in a position to appreciate the nuances: Kameshwar Wali, *Chandra: A Biography of S. Chandrasekhar* (University of Chicago Press, 1991). For a derivation of the Chandrasekhar limit, see Kerson Huang, *Statistical Mechanics* (John Wiley, 1967).

Watson and Crick's quote is from J. D. Watson and F. H. C. Crick, "A Structure for Deoxyribose Nucleic Acid," *Nature* 171, April 25, 1953, p. 737. Interestingly enough, George Gamow also played an important role as scientist and prankster in the exploration of the DNA code. See, for instance, J. D. Watson's recent book *Genes, Girls and Gamow* (Oxford University Press, 2001) and an article by G. Segrè about Gamow in *Nature* 404, March 30, 2000, p. 437.

The workings of the bombardier beetle are described in Bernd Heinrich's *Thermal Warriors* (Harvard University Press, 1996).

The quote from the *Divine Comedy,* with translation, is from Dante Alighieri's *Inferno* as translated by P. and J. Hollander, Doubleday, 2000.

Here are a few references on exotic subjects:

Solid helium: Robert Cahn, "Superdiffusion in Solid Helium," *Nature* 400, August 5, 1999, p. 512.

Helium-3: "The 1996 Nobel Prizes in Physics," news item, *Scientific American,* January 1997, p. 15.

Superfluidity: Russell Donnelly, "The Discovery of Superfluidity," *Physics Today,* July 1995, p. 30.

High-temperature superconductors: Paul Chu, "High Temperature Superconductors," *Scientific American,* September 1995, p. 162.

The Bose-Einstein condensate: Eric Cornell and Carl Weiman, "The Bose-Einstein Condensate," *Scientific American,* March 1998, p. 40.

Magnesium diboride: J. Nagamatsu et al., "Superconductivity at 39K in Magnesium Diboride," *Nature* 410, January 4, 2001, p. 63; or Charles Day, "New Conventional Superconductor Found with a Surprisingly High T," *Physics Today,* April 2001, p. 17.

Acknowledgments

WRITING A BOOK for the first time requires a certain amount of bravery, a good deal of persistence, and a large dose of faith. Luckily a number of people helped me along the way. First I would like to thank my agents, John Brockman and Katinka Matson, for encouraging me to write the book and putting me in touch with Viking, where my editor, Wendy Wolf, ably guided the book through several versions, improving style and content as well as gently prodding me along.

I have had long discussions with friends and colleagues about the book over the course of the past few years, many of them very supportive of my intention to write about topics far from the fields in which I was trained. I would particularly like to thank three of them for reading part or all of the manuscript and offering many valuable suggestions. The book is much better because of their readings. First is my colleague Philip Nelson, who generously went through the whole manuscript, pointing out awkward and occasionally just plain wrong phrases. Nick Salafsky and Peter Sterling were also of great help.

The University of Pennsylvania granted me a leave of absence for the spring term of 2001, invaluable time for completing the project, and the Rockefeller Foundation provided support for an idyllic monthlong stay at its Bellagio Study Center in June 2001.

My children and stepchildren, who for years have listened patiently to my expounding on the wonders of science, took parts of the book and read them to see if they were on target. Above all, though, my wife, Bettina Yaffe Hoerlin, provided both criticism and support. On innumerable walks, I described the topic I was planning to attack next. She forced me each time to convert the academic lecture into a story and later read and reread the various drafts of the tale that emerged, focusing each version. With love and gratitude for the help, the book is therefore dedicated to her. May our walks continue for a long time.

Index